Geofrey Kingori Gathungu
Charles Wachira Ngugi
Mwangi Mwariri

A Handbook of Organic Plants Production in Kenya

AF138642

Geofrey Kingori Gathungu
Charles Wachira Ngugi
Mwangi Mwariri

A Handbook of Organic Plants Production in Kenya

A Case of Pyrethrum Production and Utilization

LAP LAMBERT Academic Publishing

Impressum / Imprint
Bibliografische Information der Deutschen Nationalbibliothek: Die Deutsche Nationalbibliothek verzeichnet diese Publikation in der Deutschen Nationalbibliografie; detaillierte bibliografische Daten sind im Internet über http://dnb.d-nb.de abrufbar.
Alle in diesem Buch genannten Marken und Produktnamen unterliegen warenzeichen-, marken- oder patentrechtlichem Schutz bzw. sind Warenzeichen oder eingetragene Warenzeichen der jeweiligen Inhaber. Die Wiedergabe von Marken, Produktnamen, Gebrauchsnamen, Handelsnamen, Warenbezeichnungen u.s.w. in diesem Werk berechtigt auch ohne besondere Kennzeichnung nicht zu der Annahme, dass solche Namen im Sinne der Warenzeichen- und Markenschutzgesetzgebung als frei zu betrachten wären und daher von jedermann benutzt werden dürften.

Bibliographic information published by the Deutsche Nationalbibliothek: The Deutsche Nationalbibliothek lists this publication in the Deutsche Nationalbibliografie; detailed bibliographic data are available in the Internet at http://dnb.d-nb.de.
Any brand names and product names mentioned in this book are subject to trademark, brand or patent protection and are trademarks or registered trademarks of their respective holders. The use of brand names, product names, common names, trade names, product descriptions etc. even without a particular marking in this work is in no way to be construed to mean that such names may be regarded as unrestricted in respect of trademark and brand protection legislation and could thus be used by anyone.

Coverbild / Cover image: www.ingimage.com

Verlag / Publisher:
LAP LAMBERT Academic Publishing
ist ein Imprint der / is a trademark of
OmniScriptum GmbH & Co. KG
Heinrich-Böcking-Str. 6-8, 66121 Saarbrücken, Deutschland / Germany
Email: info@lap-publishing.com

Herstellung: siehe letzte Seite /
Printed at: see last page
ISBN: 978-3-659-68455-5

A HANDBOOK OF
ORGANIC PLANTS PRODUCTION IN KENYA: A CASE OF PYRETHRUM PRODUCTION AND UTILISATION

Environment friendly insecticidal plant

EDITED BY

Geofrey Kingori Gathungu, Charles Wachira Ngugi and Mwangi Mwariri

A Handbook of Organic Plants Production in Kenya

A Handbook of Organic Plants Production in Kenya: A Case of Pyrethrum, Production and Utilisation

Environment friendly insecticidal plant

Edited by Geofrey Kingori Gathungu, Charles Wachira Ngugi and Mwangi Mwariri

TABLE OF CONTENTS

LIST OF TABLES

LIST OF FIGURES

CONTRIBUTORS

Geofrey K. Gathungu. Department of plant sciences, Faculty of Agriculture and Environmental Studies, Chuka University, P.O. Box 100-60400, Chuka, Kenya. gkgathungu@yahoo.com

Charles W. Ngugi. Kenya Agriculture and Livestock Research Organization. Beef Research Institute, Lanet. P.O. Box 3840-20100, Nakuru, Kenya. cwngugimolo@gmail.com

Mwangi Mwariri. Kenya Agriculture and Livestock Research Organization. Headquarters, P.O. Box 57811-00200, Nairobi, Kenya. mmwariri@yahoo.com

FOREWARD

Agricultural productivity is limited by numerous biotic and abiotic constraints, but soil fertility and pest management are critical. Commercial fertilizers and pesticides have limited access in rural areas due to their high cost and resource-poor farmers cannot afford them. They have also become ineffective due to soil polarization and pest resistance. In the recent past several botanical plants have been found to be an effective alternative to soil fertility improvement and pest control. Promotion of the botanical plants would have enormous impact on a farmer's ability to manage soil resource and pests as they are promising alternative to synthetics towards organic crop production and pest control. Various crude products are being used by farmers but there is limited information available on application, efficacy, toxicity, persistence, shelf life and safety of these products. Several farmers have used plants in this way, making the technology familiar, trusted and acceptable, but their priority in agricultural policy is low. The optimization of their full potential, especially for the poorest farmers, is constrained by inadequate product evaluation and development which, if improved, will increase the reliable options available to farmers. Exploitation and optimal use of botanical plants can be enhanced through development and dissemination of knowledge and capacity building for research and initiation of innovations geared towards their sustainable development and use in organic crop production. Botanical plants can be used to produce biopesticides at the farm level. Biopesticide registration guidelines and policy recommendations require to be developed and formulated and documents produced to facilitate extraction and use of the same at the farm level to facilitate their uptake and safe use. Biopesticides use due to their low residue levels will go a long way towards minimization of hazards related to use of synthetic pesticides. Organic production of crops is of critical importance to the resource poor farmers as it increases on their value and competitiveness in markets. There is need to strengthen the capacity of farmers to exploit botanical plants and optimize their use in crop production. Botanical plants can also provide marketable products for farmers and their commercialization will provide both an additional income stream to poor farming communities. This will reduce the high level of rural poverty by making agriculture more competitive, increasing food security and raising poor farmers' incomes by cost effectively increasing crop yields, and reducing post harvest losses. Some of the botanical plants include pyrethrum, neem, *Tephrosia vogelii*, Ocimum, Marigold, *Tithonia diversifolia, Crotalaria juncea* , Datura, black jack among others. However, today except for pyrethrum there is no coordinated cultivation or conservation of most of the botanical plants and most of them are sourced from the wild. Exploitation of botanical plants can be achieved through sustainable production of botanical pesticides through improved propagation, cultivation, germplasm conservation and harvesting. There is also need to invest in product development, formulation standardization and application from the simple crude extracts to the refined active ingredients. Pyrethrum is the only botanical crop which has organized production system and is well commercialized. Pyrethrum production has been constrained by lack of knowledge and information on production. This handbook will impact greatly on production and utilization of pyrethrum on areas where it is grown because it details the principles and practices required in the production value chain. In my current teaching, training and research career there is a lot of emphasis on organic crop production and therefore this handbook will be resourceful to farmers, students, researchers and stakeholders in agriculture development.

GATHUNGU GEOFREY KINGORI
Lecturer, Agronomy and Horticultural Crops Production
Chuka University, Kenya

CHAPTER ONE

1.0. ORGANIC CROP PRODUCTION

1.1. BACKGROUND

The world's climate is changing fast, and will continue to do so for the foreseeable future, no matter what measures are now taken. Temperatures are rising, rainfall is increasing in some areas and declining in others, seasonal patterns and pest and disease distribution are changing, and extreme weather events are becoming more frequent and severe, posing new risks to food and farming. There is need therefore to develop means or strategies by which farming systems can adapt to cope with the changes, as well as the potential of agriculture to mitigate our climate impact. This recognition has led to the concept of 'climate-smart agriculture'. The agricultural sector is the backbone of Kenya's economy and the means of livelihood for most of our rural population. Sustained agricultural growth is critical to uplifting the living standards of our people as well as generating rapid economic growth. Agricultural sector is a key driver in achieving the 10 per cent annual economic growth rate envisaged under the economic pillar of Vision 2030. Agriculture is the mainstay of the Kenyan economy directly contributing 26 per cent of the GDP annually, and another 25 per cent indirectly. The sector accounts for 65 per cent of Kenya's total exports and provides more than 70 per cent of informal employment in the rural areas. The agricultural sector comprises the following subsectors: crops, livestock, fisheries, land, water, cooperatives, environment, regional development and forestry.

Agriculture plays a vital role in a developing country in fulfilling the food requirement of the growing population and it also plays a role in improving economy of the country. Agricultural productivity is limited by numerous biotic and abiotic constraints. Soil fertility and pest management are the most important. Key practices include alternate wetting and drying (AWD) of the soil during crop growth rather than continuous flooding/sprinkler irrigation, application of organic fertilizers, such as manure, and reduced use of inorganic fertilizers and pesticides. Commercial fertilizers and pesticides usually work but have limited distribution in rural areas and have become ineffective due to soil polarization and pest resistance. The Green Revolution technology adoption where inputs like fertilisers, pesticides were used between 1960 to 2000 increased wide varieties of agricultural crop yield per hectare which increased 12-13% food supply in developing countries. However, generally in the world over, food insecurity and poverty still prevails prominently and the use of chemical biopesticides and fertilisers have caused negative impact on environment by affecting soil fertility, water hardness, development of insect resistance, genetic variation in plants, increase in toxic residue through food chain and animal feed thus increasing health problems and many more. From these challenges it has made it essential to introduce measures which can counteract the said effects.

Organic crop production which includes use of botanical pesticides and fertilizers (Biopesticides and Biofertilizers) can play a major role in dealing with these challenges in a sustainable and friendly way. This farming system respects the biological relationship that exists in nature thereby fostering natural resource and environmental conservation. Organically grown produce is therefore healthy as it is grown without use of any (or limited) chemical fertilizers and synthetic pesticides, additives and preservatives and without genetic manipulation of living organisms as happens with genetically modified foods.Organic agriculture is an ecological production management system that promotes and enhances biodiversity, biological cycles, and soil biological activity. It integrates cultural, biological, and mechanical practices that foster cycling of resources, promote ecological balance, and conserve biological diversity. Botanical plants (BPs) are an effective alternative to soil fertility improvement and pest control both in crop and livestock production, and public health. Soil health is the foundation of organic farming systems as it provides essential nutrients to plants and supports a diverse and active biotic community that helps the soil resist environmental degradation.

Botanical plants can be used as fallow, cover or rotation crops, green or for compost manures. Promotion of (BPs) would have enormous impact on a farmer's ability to manage soil resource and pests which are the basis of increased horticultural production. Extracts from BPs could have important roles in biological management strategies for control of pests and soil fertility improvement. Botanical plants are plants whose extracts have antimicrobial, insecticidal or repellent activities or are rich in nutrients. BPs are natural products derived from plant roots, leaves, berries, barks, stems and flowers and they represent an environmentally friendly alternative for crop protection and soil improvement. BPs are one of the most adequate inputs for Organic Farming and Integrated Pest Management.

There has been growing pressure to reduce residue levels from conventional chemical pesticides due to regulations on changes in MRLs and retailers to minimise detectable residues. This will assist to develop sustainable 'pest' control systems that maintain environmental welfare, crop quality, productivity and profitability. Integrated Crop Management (ICM) will remain the only way for reduced chemical pesticide use and residue levels. Some of the main rationales behind organic farming are the production of food of a better quality and the protection of the soil and the environment. With organic farming the aim is to produce food of a better quality and to develop sustainable crop production systems, which protect our soil and our environment. BPs are a promising alternative to synthetics towards organic horticultural farming practices. The area under organic crop cultivation may rise because of the growing demand of organic food, as a result of increasing health consciousness among the people.

1.2. ORGANIC AGRICULTURE PRODUCTION SYSTEM

Organic agriculture includes all agricultural systems that promote the environmentally, socially and economically sound production of food and fibers. Organic farming relies on techniques such as crop rotation, green manure, compost and biological pest control. It uses fertilizers and pesticides but excludes or strictly limits the use of manufactured (synthetic) fertilizer, pesticides (herbicides, insecticides, and fungicides), plant growth regulators (hormones), livestock antibiotics, food additives, GMOs among others. Organic agricultural methods are internationally regulated and legally enforced by many nations, based in large part on the standards set by the International Federation of Organic Agriculture Movement (IFOAM), an international umbrella organization for organic farming organizations established in 1972.These systems take local soil fertility as a key to successful production. Organic agriculture dramatically reduces external inputs by refraining from the use of chemo-synthetic fertilisers, pesticides, and pharmaceuticals. Instead it allows the powerful laws of nature to increase both agricultural yields and disease resistance. In organic agriculture as a production system, the farmer should combine cultural (e.g., planting disease-resistant varieties); biological (e.g., pheromone traps); and mechanical (e.g., appropriate tillage) techniques into a productive management system that minimizes the impact to the off-farm environment. Biopesticides are biochemical pesticides that are naturally occurring and are living organisms (natural enemies) or their products (phytochemicals, microbial products) or byproducts (semiochemicals) which can be used for the management of pests that are injurious to plants by nontoxic mechanisms. The most commonly used biopesticides are living organisms, like biofungicides

(Trichoderma), bioherbicides (Phytopthora) and bioinsecticides (Bacillus thuringiensis) including incorporation of Bt gene into the plant genome so that the transgenic plant synthesizes its own substance that destroys the targeted pest. Biopesticides has several advantages including inherently less harmful and less environmental load, they affect only one specific pest or, few target organisms, are effective in very small quantities and often decompose quickly, resulting in lower exposures and largely avoiding the pollution problems and when used as a component of Integrated Pest Management (IPM) programs, they contribute greatly.

Use of botanicals as sources of biopesticides and biofertilizers in crop production is now emerging as one of the important means towards pest control and soil fertility management due to increasing pesticidal pollution. There is rich traditional knowledge base available within the highly diverse farming communities in Kenya that may provide valuable clues for developing newer and effective biopesticide for increased adoption by the farmers. Organic agriculture should be based on the principle of sustainability where the farmer meets the needs of the present without compromising the ability of future generations to meet their own needs. It is important to focus on building the soil with farm-generated fertility. Converting from conventional to organic systems over time, add organic matter, populations of soil microbes and soil invertebrates resulting in a rich, healthy, biologically active soil that produce healthy plants. In return also the farmers need fewer off-farm inputs because their crops get better adapted to resist drought, diseases, and insects. Successful organic farmers focus on preventing problems, rather than reacting to them like where they prevent insect problems by providing habitat for beneficial insects that keeps populations of harmful insects in check. Although farmers are somehow aware of the importance of using botanical plant products as alternatives to chemical pesticides, their widespread use of these plant products is limited. There exists various challenges that have hindered exploitation of the botanical crops in organic agriculture (table 1). Currently there is lack of reliable knowledge of botanical plants sowing techniques, plant propagation techniques, pest and disease management/preventive and promotive care to build up disease resistance and their cultivation. Extracts of plants like neem, garlic, onion, turmeric, ginger, tobacco, papaya, aloe, custard apple etc and their effects on curing plant diseases have not been well tested. Their processing and provision in usable form to farmers will be one way of popularizing their acceptance and use.

1.2.1. Types of Biopesticides

Biopesticides fall into three major classes that includes Microbial pesticides, Plant pesticides and Biochemical pesticides .

Microbial pesticides consist of a microorganism (e.g., a bacterium, fungus, virus, or protozoan) as the active ingredient. Microbial pesticides can control many different kinds of pests, although they are specific to target pest[s] (some fungi control certain weeds, and other fungi kill specific insects). The most widely used microbial pesticides are Bacillus thuringiensis, or Bt bacterium. While some Bt's control moth larvae found on plants, other Bt's are specific for larvae of flies and mosquitoes.

Plant pesticides are pesticidal substances that plants produce from genetic material that has been added to the plant. For example, gene for the Bt pesticidal protein can be introduced into the plant's own genetic material. Then the plant, instead of the Bt bacterium, manufactures the substance that destroys the pest.

Biochemical pesticides are naturally occurring substances that control pests by non-toxic mechanisms. Biochemical pesticides include substances, such as insect sex pheromones, that interfere with mating, as well as various scented plant extracts that attract insect pests to traps. If farmers can adopt the utilization of biopesticides which are generated from natural botanical plants they can realize various advantages. Biopesticides usually are inherently less harmful than conventional pesticides, they affect only the target pest and closely related organisms, in contrast to broad-spectrum conventional pesticides that may affect organisms as different as birds, insects, and mammals , they are effective in very small quantities and often decompose quickly, thereby resulting in lower exposures and largely avoiding the pollution problems caused by conventional pesticides, and when used as a component of Integrated Pest Management (IPM) programs, they can greatly decrease the use of conventional pesticides, while crop yields remain high.

1.3. WHY USE BOTANICAL PLANTS

Botanical plants extracts and phytochemicals have long been a subject of research in an effort to develop alternatives to the conventional pesticides and fertilizers. Use of the organic manures, amendments, plant extracts and bio fertilizers is gaining considerable attention in organic farming because of being safe to use compared to synthetic pesticides, cheap and affordable for small-scale farmers and are not very persistent and most of them break down hence don't have a long lasting contaminating effect on the environment. Towards food safety-agricultural pest management, botanical pesticides are best suited for use in organic food production. Climate change has also led to declining soil fertility and changes in pest dynamics.

1.4. MODE OF ACTION OF BOTANICAL PLANTS

BP extracts interfere with the growth where on contact they act as neuro-muscular toxin or interfere with mitochondrial activity. On feeding they act as stomach poison and also interfere with reproduction - mating disruption, monitoring or attract and kill strategies. Other botanical extracts act as antifeedants for many pest species with no deleterious effects on humans, animals or beneficial insects.

1.5. THE PRINCIPLE AIMS OF ORGANIC CROP PRODUCTION

Organic farming does not mean going 'back' to traditional methods. Many of the farming methods used in the past are still useful today. Organic farming takes the best of these and combines them with modern scientific knowledge. Organic farmers do not leave their farms to be taken over by nature; they use all the knowledge, techniques and materials available to work with nature. In this way the farmer creates a healthy balance between nature and farming, where crops and animals can grow and thrive. Therefore the aim of organic crop production should include to;

- Produce food of high nutritional quality in sufficient quantities.
- Interact in a constructive and life-enhancing way with natural systems and cycles.
- Encourage and enhance biological cycles within the farming system, involving micro-organisms, soil flora and fauna, plants and animals.
- Maintain and increase long-term fertility of soils.
- Promote the healthy use and proper care of water, water resources and all life therein.
- Help in the conservation of soil and water

- Use as far as possible renewable resources in locally organized agricultural systems
- Work, as far as possible, within a closedsystem with regard to organic matter and nutrient elements
- Work, as far as possible, with materials and substances which can be reused or recycled, either on the farm or elsewhere
- Give all livestock conditions of life which allow them to perform the basic aspects of their innate behavior.
- Minimize all forms of pollution that may result from agricultural practice
- Maintain the genetic diversity of the agricultural system and its surroundings, including the protection of plant and wildlife habitats.

Table 1: Challenges of botanical plants

Category	Challenges
Plant species	• Species available not clearly known
Planting materials	• Shortage of seedlings
	• Use of low yielding seedlings
Crop husbandry	• Low farmer knowledge on improved production technology
	• Low input use
Harvesting	• Harvesting stage not well defined
Processing	• Protocols not well developed
Utilisation	• Low or non existent in some cases
Markets	• Poor /unhygienic market facilities
	• No organized marketing leading to farmer exploitation
	• Lack of market information
	• Price instability
	• Failure to meet market quality standards
Customer awareness	• Awareness lacking

1.5.1. Challenges of modern, intensive agriculture

Intensive agriculture involves heavy capital investment mainly in the use of synthetic pesticides and fertilizers to maintain high yield per unit area. This results to various negative effects to man and the environment which includes;

- Water pollution -Artificial fertilizers and herbicides are easily washed from the soil and pollute rivers, lakes and water courses.
- The prolonged use of artificial fertilizers results in soils with a low organic matter content and which is easily eroded by wind and rain.

- Dependency on fertilizers. Greater amounts are needed every year to produce the same yields of crops.
- Artificial pesticides can stay in the soil for a long time and enter the food chain causing health problems to both man and livestock.
- Artificial chemicals destroy soil micro-organisms resulting in poor soil structure and aeration and decreasing nutrient availability.
- Pests and diseases become more difficult to control as they become resistant to artificial pesticides.
- The numbers of natural enemies decrease because of pesticide use and habitat loss.

1.5.2. Benefits of organic farming

Organic farming provides long-term benefits to people and the environment that includes;
- Increase long-term soil fertility
- Control pests and diseases without harming the environment
- Ensure water quality (stays clean and safe)
- Use resources which the farmer already has, so the farmer needs less money to buy farm inputs. Low cost of production
- Produce nutritious food, feed for animals and high quality crops to sell at a good price.

1.6. DRAWBACKS/BARRIERS TO BOTANICAL PLANTS COMMERCIALIZATION

a) Availability and sustainability of the botanical resource. Most of BPs except pyrethrum and neem are not obtainable on an agricultural scale. Their cultivation has not been commercialized.

b) Lack of validation, standardization and safety assessment of chemically complex extracts (active ingredient). Currently analytical method and the equipment necessary for analysis and storage facilities not readily available. There is no production infrastructure.

c) Lack of guidelines, policies and regulatory framework. This remains a barrier to the commercialization of new BPs as there is no distinction between synthetic pesticides and biopesticides due to high regulatory costs

Currently except for pyrethrum there is no coordinated cultivation or conservation of most of the BPs and most of them are sourced from the wild. Exploitation of BPs will only be achieved with sustainable production through improved propagation, cultivation, germplasm conservation and harvesting. There is need to invest in product development, formulation standardization and application from the simple crude extracts to the refined active ingredients. There is therefore need for;
- Developing production (breeding, agronomic, crop protection) and post harvest technologies
- Validation of their efficacy and safety and standardization of the formulations (Value addition to improve on profitability)
- Increase production and utilization awareness through dissemination of developed technologies
- Develop policy and legal framework on BPs
- Partnership and networking

1.7. WAYS OF PROMOTING BOTANICAL PLANTS UTILIZATION IN KENYA

a) **Legalization:** The pesticide legislation protocol in Kenya as spelled out by the he Pest Control Act covers only chemical pesticides and treats botanical plants (biopesticides) products as similar products thereby inhibiting their commercialization and use. Regulatory approval has become so costly and time consuming that only multinational agrochemical companies have the resources to satisfy regulatory requirements for their insecticides creating less room for community based biopesticide commercialization to see widespread agricultural use

b) **Awareness:** In rural areas, the main concern is the indiscriminate use of pesticides. Most of the populations using synthetics are ignorant of their dangers. There is cross misuse of pesticides, within the farming communities. The government should protect the farmers by promoting safe procedures for handling and storing agrochemicals. There is need to carry out, in collaboration with local farmers, demonstration projects and to develop technologies for management of traditional and non-chemical methods of pest control. There is need to further work with local farmers, to evaluate the effectiveness of traditional and non chemical methods alone or in integrated pest management (IPM) strategy.

c) **Favourable policy**: Favourable government policy support will enable variable stakeholders who promote the use of biopesticide products achieve much in terms of product development, legality and promotion. The registration, restriction and banning of pesticide products, as well as the development and commercialization of biopesticides is all governed by state policy.

d) **Funding**: Using biopesticides is a lot more expensive than dealing with synthetics. There is need for positive support through funding for research institutions, to facilitate mass production and commercialization of biopesticides. This will improve on their availability.

Botanicals are popular among the illiterate and resource poor households. To intervene, scientists need to upgrade local botanical pesticides into marketable products that can attract all cadres of farm households. The positive effects of age indicate the prevalence of botanicals among the old, who struggle to preserve local knowledge and practices. Therefore, gender, literacy levels, wealth endowments in form of land and old age are key factors to be considered for intervention for botanical pest controls in the field and stores.

In this production handbook principles and practices of pyrethrum will be widely described as it has an established industry and is also widely grown in Kenya.

CHAPTER TWO

2.0. PYRETHRUM PRODUCTION IN KENYA

2.1. INTRODUCTION

Pyrethrum was introduced in Kenya in 1928. Pyrethrum flowers contain pyrethrins which are used in the manufacture of natural insecticides. It is among the major foreign exchange earners, ranking fourth after tea, horticulture and coffee. The industry provides income directly to about 250,000 small scale farmers. Pyrethrum is the main cash crop in the areas where it is grown. The industry provides employment directly or indirectly to about 1.5 million workers. Pyrethrum is a labour intensive crop and operations such as planting, weeding, picking and drying are carried out manually.

Kenya has been the world's largest producer and exporter of pyrethrum for over 80 years accounting for between 60-70% of the global supply. The pyrethrum flower contains insecticidal compound whose extract is used in production of most widely used botanical insecticides. Efforts have been made to introduce pyrethrum in other countries in Africa, Asia and South America without much success due to lack of ideal growing conditions like temperate weather, moderate well distributed rainfall, low night temperatures, rich well drained volcanic soils and high altitudes from 1800 metres thus leaving Kenya with the largest market share. The other producers are Tasmania, Tanzania, Rwanda, Uganda and Papua New Guinea. The current demand for pyrethrum in the world market is about 20,000 metric tons of dried flowers. Despite the large world market Kenya's production has declined from 18,900 tons of dry flowers in 1981-1982 to 2205 tons in 2004/05, to 776 tons in 2007/08 and to 518 tons in 2010/011 and the production has been constantly decreasing. The decline is attributed to climate change that favours drought and high temperatures, competition with food crops, shortage of labour, inadequate clean planting material, diseases and pests, high cost of inputs, lack of credit, inadequate extension services, low pricing and non-payments.

In Kenya, climatic conditions suitable for growing pyrethrum are found in Rift Valley, Central and Nyanza provinces with some insignificant production in Eastern and Western provinces. The major growing counties are Nakuru, Nyandarua, Kisii, Nyamira and Uasin Gishu accounting for close to 85% of the total national production. Other counties where pyrethrum is grown includes Bomet, Baringo, Narok, Keiyo Marakwet, Pokot, Nandi, Kiambu, Nyeri, Meru, Laikipia, Kericho, Trans Nzoia and Mt. Elgon. Kenya being the leading producer of pyrethrum extract, the crop is an important foreign exchange earner and a significant contributor to the economy.

2.2. ORIGIN OF PYRETHRUM

The importance of pyrethrum as a useful botanical insecticide in control of domestic pests was first recognised and exploited in Persia around 1800. Later, pyrethrum plants were grown and flowers commercially sold as an insecticide powder in Europe in 1920s. The crop spread to China through present day Iran and was later introduced and commercially cultivated in Japan, which became the main producer between 1918 and 1945. The pyrethrum species used during this period were mainly *Chrysanthemum roseum* and *C. corneum,* which had low pyrethrins concentration.

Pyrethrum was introduced in Europe 1840s. Dalmatia (Yugoslavia) became the leading world producer until the end of World War I. The crop was sparsely grown in other parts of Europe including England. It was introduced in Kenya from Dalmatia and England in 1928-1929 and in Rwanda, Uganda, Tanzania and Zaire about the same time. The species introduced was *C. cinerariaefolium* which has high flower yield, pyrethrins concentration and superior growth characteristics. Attempts to introduce pyrethrum on a commercial basis in other parts of the world including Equador, India and Nepal have been hampered by economic viability. Although other major producing countries such as Tanzania, Rwanda, Tasmania, Uganda, Papua New Guinea and Zaire produce pyrethrum, the industry in those countries is not enough to meet the world demand. Significant pyrethrum production has been achieved in Tasmania and Papua New Guinea.

Twenty years after the introduction, Kenya became the most important source of pyrethrum in the world because of suitable climatic conditions, altitude and soils, development of high yielding cultivars with high pyrethrins concentration, well organised smallholder production system, and ready market for natural insecticide products in the world market. Kenya started exporting dry pyrethrum flowers and powder in 1933. With the introduction of efficient pyrethrin extraction methods, Kenya now exports pyrethrum in form of a standardised extract.In Kenya, pyrethrum is cultivated almost entirely by small-scale farmers on an average of ½ acre of land. However, few large scale farmers grow the crop on between 20 – 50 acres. In both cases mechanization is basically at land preparation and all other operations are done manually. One hectare accommodates 52,000 plants producing about 1,000kg of dried pyrethrum flowers annually. This quantity yields about 25kg of highly refined extract. Ready flowers are picked at intervals of two weeks with picking continuing for nearly a year from July to April. Although pyrethrum is a perennial crop, a typical plantation lasts for three to four years. Harvested flower heads are sun-dried and delivered through rural co-operatives to the centrally located factory of the Pyrethrum Board of Kenya (PBK). Farmers are paid according to the weight and content of their harvests. The growing demand for "organic" and "natural" pesticides with minimal residue limits (MRLs) has increased international demand for pyrethrum, despite the existence of synthetic chemical substitutes.

In the past, pyrethrum production was only valued by small-scale Kenyan farmers as a way of raising some income for their household livelihood. All the dry flowers produced was sold to Pyrethrum Board of Kenya, the processing industry and few farmers were really aware of the potentials of pyrethrum at the farm level. Recent developments, however, have opened up new and interesting options for pyrethrum use in pest and soil management and livestock nutrition on smallholdings. Pyrethrum powder is highly effective in preventing insect damage to stored grain of maize, wheat, barley and oats. If the powder, is mixed with dry grain immediately after harvest, it controls weevils, beetles, grain borers and meal worms for up to two years. The 'waste materials or powder' of the dried pyrethrum flowers processing that remains after pyrethrins have been extracted is referred to as the pyrethrum marc. It has been found to be a healthy feed supplement for livestock, it reduces load of intestinal parasites and ticks occurrence, and an improved general animal appearance. Pyrethrum marc has in other areas been applied in the funnel of maize to control maize stem-borers, one of the most important pests of Kenya's staple food. Extract from locally dried flowers when mixed with deep frying oil or milking serve has been used to control termites and ticks respectively.

2.3. PYRETHRINS

Pyrethrum is a small herbaceous perennial crop, commercially grown for its highly valued pyrethrins found in its flowers. The pyrethrin is a term used to describe 6 active ingredients used in the formulation of various environment friendly insecticides. The natural pyrethrins have unique qualities of an ideal, all-purpose pest

19

control agent with little development of resistant strains. Pyrethrins have a flushing effect and have an effective rapid knockdown effect on a wide variety of insect pests. They have low toxicity to mammals and other warm-blooded animals. Their repellence effect is important than the killing effect when protecting food. Pyrethrins are non-inflammable and are highly biodegradable leaving no residues after use. Therefore, pyrethrins being natural provides a level of safety and environmental comforts to today's general public concern on health issues. These inherent properties permit its use in domestic premises, factories, hospitals, storage facilities, water systems, fish ponds, and on humans, livestock and pets.

As a result of environmental legislation, increased pests resistance to synthetic pesticides, growing awareness amongst consumers and industrial research and development and the high cost of chemical insecticides, interest in safe natural insecticide use is expanding. Extra impetus in research in Kenya and other parts of the world to increase stability of pesticide formulation will enable the use of pyrethrins formulations as an all-purpose crop protection and disease vector control agent both indoor and outdoor. The demand for pyrethrum in the world is about 20,000-25,000 ton. Due to the current global health concerns due to concentration of chemical residues in agricultural products more countries are imposing strict legislation rules on use of synthetic pesticides. This in the future will continue to increase the demand for pyrethrum.

2.4. MORPHOLOGICAL CHARACTERISTICS

Pyrethrum is a perennial herbaceous daisy plant that is propagated either through seed, vegetative splits or tissue-culture. The pyrethrum flower head consists of several small florets aggregated together on a convex receptacle. There are 2 kinds of florets. The ray florets have white petals which form an outer ring of the flower head and the yellow disc florets forming the inner and most of the flower head. They are female with no stamen for pollen production while the disc florets are bisexual with both female and male parts. The pyrethrins occur as viscous oil in oil glands mainly in the ovaries of the disc florets.

Pyrethrum plant produces long branched tillers, which grow from the crown. The tillers barely grow over 75 cm and have lobed leaves and daisy flowers at the end of the long flower stalks. The plant has a fibrous root system, which grows and extends mainly in the top 30cm of the soil profile. Pyrethrum is established through seeds, splits of mature plants or tissue cultured seedlings. In Kenya pyrethrum is mainly propagated through splits for the clones and seed for the varieties. Tissue culture is mainly used to raise initial clean propagation materials that is multiplied in nurseries for future sale to farmers. Farmers in return renew their fields after 3-4 years through splitting.. Each of these methods of propagation has its own advantages and disadvantages. Propagation through seeds is easy but matures after one year while propagation through splits takes 3 months to flower and gives a uniform population. Rapid multiplication through tissues is fast and efficient but requires heavy capital investment.

2.5. BOTANY

2.5.1. The Flower

The pyrethrum flower is typical of the compositae family, and is a collection of white relatively small flowers (florets) which are arranged on a convex receptacle. The yellow disc florets are located at the centre of the receptacle and are surrounded by an outer ring of white ray florets (figure 1).

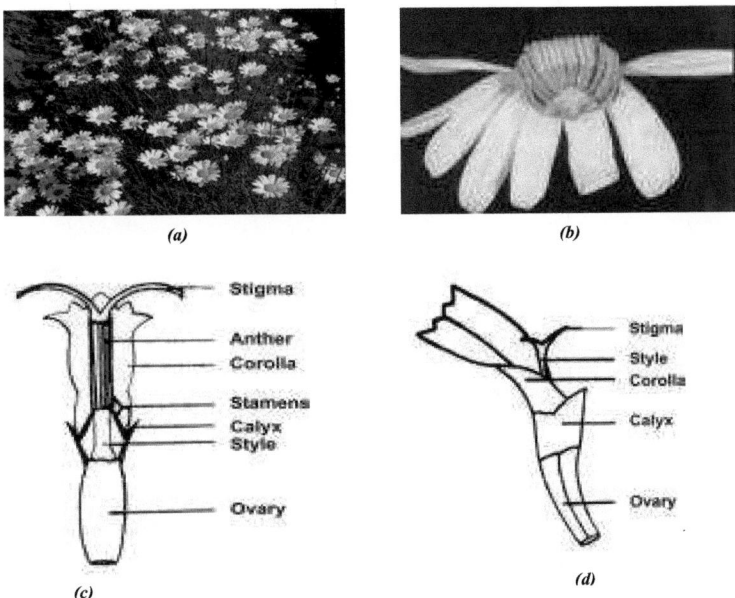

(a) *(b)*

(c) *(d)*

Figure 1: a) Tillering plant b) Vertical section of flower head c) Disc floret d) Ray floret

The ray florets differ in basic structure from the disc florets. The corolla is enlarged to form the conspicuous white petals of the flower head. The style projects through the folded base of the corolla and its bi-lobed stigma has the same structure as that of the disc florets. The ray florets have only the female parts.

Each of the disc florets has a yellow, tubular corolla with a small calyx at its base. Five stamens rise from the base of the inner surface of the corolla and end at the elongated anther. The style is at the centre of the floret, which rises from the ovary. The style ends in a bi-lobed stigma. The outer disc florets open 1st as the flower starts to mature and progressively proceeds opening towards the centre of the flowers. The disc florets have both the female and male parts.

2.5.2. Pollination

Pyrethrum is cross-pollinated by insects, especially bees. The ray florets open from the outer row of the disc florets towards the centre. Several florets open and shed their pollen every day. The time taken from the opening of the first disc florets until all the florets are fully open or overblown is 14-15 days depending on weather and type of cultivar. Self-pollination does not occur due to self-incompatibility. Warm dry spells and many different insects' speed the pollination process. Depending on the weather, it takes between 6-8 weeks for the just opened flower to reach the mature seed stage.

Insect pollination

Figure 2: insect involvement in pyrethrum flower pollination

2.5.3. Flower development

The pyrethrum flower passes through 8 stages of development from bud formation to the seed stage (table 1, figure 3).

2.5.4. Flower periodicity

Fresh flower yields show a marked periodicity in a year. Such flower production is associated with rainfall patterns and sunshine spells. Flower flushes occur immediately after the long rains in March - April and low temperatures of July and August and also after the short rains in October and November.

Flower size, dry matter content and pyrethrins content vary depending on rainfall. During the rainy spell, the flower size enlarges, the dry matter content diminishes and the pyrethrin content increases. During the dry spell, the reverse occurs. Most pyrethrum clones show this pattern, although the extent may differ slightly among the clones.

2.5.5. PYRETHRIN CHEMISTRY

Pyrethrum (*Chrysanthemum cinerariefolium* Vis) is one of the few plant species that is commercially cultivated for its active natural pyrethrins which are used in formulation of insecticidal products. The insecticidal constituents of the pyrethrum flowers are collectively called the pyrethrins.

The pyrethrins are the collective term used for the 6 related chemical esters, pyrethrin I, cinerin I, jasmolin I and pyrethrin II, cinerin II, jasmolin II, which have extraordinary insecticidal properties. They are mainly concentrated in the achenes of the flower heads. Although there is variation in the proportion of each component of pyrethrins in the flower head, pyrethrins I and II comprise the highest proportion of 73.0%, cinerin I and II have 19.0% while jasmolin I and II have 8.0%. Pyrethrin I, cinerin I and jasmolin I comprise the pyrethrin I fraction while pyrethrin II, cinerin II and jasmolin II comprise the pyrethrin II fraction (Table 2). The pyrethrin II fraction is derived from the pyrethric acid while the pyrethrin I fraction is derived from the chrysanthemic acid.

Table 2: Stages of pyrethrum flower development

Stage	Description	Approximate pyrethrins content (%w/w)
Stage I	A well-developed closed bud	1.36
Stage II	Ray florets vertical	1.60
Stage III	Ray florets horizontal, the 1st row of disc florets open	1.64
Stage IV	Inflorescence with about 3 rows of the disc florets open	1.73
Stage V	An inflorescence with nearly all the disc florets open	1.79
Stage VI	Early overblown condition; the colour of the disc florets diminishing but the ray florets still intact	1.70
Stage VII	Late overblown condition, little colour remaining but the disc florets still intact, ray florets now dried out	1.57
Stage VIII	Disc florets fallen, stem dry 2 cm below head-suitable for collection for seed	1.29

Table 3: Composition of pyrethrins

Acid	Keto-Alcohol	Pyrethrin Fraction
Chrysanthemic	Pyrethrin I, Cinerin I, Jasmolin I	Pyrethrin I
Pyrethric	Pyrethrin II, Cinerin II, Jasmolin III	Pyrethrin II

The ratio of pyrethrin I to pyrethrin II, referred to as the PI/PII is 1.2 - 1.5 in Kenyan flowers. Pyrethrum is in the chrysanthemum genus and belongs to the compositae family. *Chrysanthemum cinerariefolium* form the basis of the pyrethrum industry in Kenya. The pyrethrum plant is the source of naturally occurring insecticide called pyrethrins which are lethal to insect pests and are relatively safe to humans and other mammalian species, they are readily biodegradable and therefore, does not accumulate in the environment. Pyrethrum based insecticides are soft, ecologically friendly and have been used worldwide in and around human dwellings, in food handling establishments, on domestic animals, pets and public health utilities.

There are over 10,000 insect species which influence mankind adversely. Although 6 billion USD is used to control insects they still consume about 30-40% of the food grown. Diptera are vectors of several diseases such as malaria, filariasis, schistosomiasis, trypanosomiasis and dysentery which cause suffering or death to people and livestock. Other pests such as lice, fleas, bugs and ticks cause irritation and spread diseases.

(a)   (b)   (c)

(d)   (e)   (f)

Figure 3: floral development stages (a) bud stage (b) ray florets vertical (c) ray florets horizontal with 1 row of disc florets open (d) ray florets horizontal with 3-4 rows of disc florets open (e) overblown condition (f) head-suitable for seed collection

2.5.6. DISTRIBUTION OF PYRETHRIN IN THE PLANT

Pyrethrins are present almost in the whole of the pyrethrum plant. However, it is only the flower heads that contain pyrethrin for economic extraction (table 4). The pyrethrin, identified as resin, occur in the plant tissue as droplets in the cells, in the intercellular spaces and in duct lines with a thin walled epithelium. About 94% of the pyrethrin occurs in the achene and ovaries. The insecticidal pyrethrin is the fluid in the secretory system of the florets. The inner walls of the achene are lined with several layers of secretory tissue in the ribs of other achene. Pyrethrum flowers with high pyrethrins content have a thicker layer of secretory tissue and larger intercellular cavities filled with the resin.

Table 4: Composition of pyrethrin in plant parts

Plant parts	Pyrethrin distribution	Pyrethrin (%)
Oil gland	1.7	54.7
Crown	0.60	19.3
Style	0.50	16.1
Leaves and petals	0.182	5.9
Roots	0.087	2.8
Stems	0.036	1.2

CHAPTER THREE

3.0. PYRETHRUM GROWTH AND DEVELOPMENT REQUIREMENTS

3.1. SOILS

Pyrethrum grows on different soil types but thrives best on fertile, deep and well-drained loamy volcanic soils which have good texture and structure. Such soils are rich in nutrients and have good water holding and infiltration capacity. The crop's performance is poor in poorly drained soils due to poor root establishment. The plants should be planted on ridges in areas with excessive rainfall and poor drainage.

The soil pH should not be below 5.6. Flower yields are low in acidic or alkaline soils. Acidic soils have a phosphorus deficiency resulting from P-fixation due to high levels of aluminium. Farmers are therefore advised to use adequate amounts of phosphate fertilisers, gypsum or lime (100 kg ha^{-1}). Alkaline soils can be amended by applying sulphur based acidic fertilisers to lower the soil PH. Alkaline soils have low water filtration rates and unfavourable physical condition.

Application of these soil amendments improves the soil reaction and physical properties which in turn improve the availability of essential elements. They also improve the activities of soil organisms that enhance decomposition of organic matter, fixation of atmospheric nitrogen and mineralisation of nutrients.

3.2. RAINFALL

The crop thrives best in areas that receive more than 760mm annual rainfall evenly distributed throughout the year. A monthly rainfall distribution of about 100 mm is ideal for continued flower production. Flushes of flowers are produced after some rainfall. A total amount of 1000 mm is adequate to sustain growth. In areas where temperatures are usually very high such as Kisii more than 1500 mm is needed to compensate for the rapid evapotranspiration. However, in areas with cloudy and misty conditions where evapotranspiration is low, a precipitation of 750-1000mm is adequate for cultivation. Short sunny spells during the rainy season promote flushes of flower production. If the dry spell is longer than 3 months, flower production is greatly reduced. The amount and distribution of rainfall in Kenya depend on the area and the year resulting to varied rainfall patterns (Table 5). On the basis of rainfall patterns, the pyrethrum growing zones may be divided into 3 major regions.

3.2.1. Nyanza Region

The pyrethrum growing region is mainly the highlands of Kisii, Gucha and Nyamira districts in south Nyanza. The climate is mainly wet with continuous rainfall. The area is situated at the centre of the convergence zone of the lake winds and easterlies. Pyrethrum flower production occurs throughout the year.

3.2.2. Rift Valley

The pyrethrum growing districts are Uasin Gishu, Keiyo Marakwet, Trans-Nzoia and Nandi in the North rift and Nakuru, Bomet, Kericho, Baringo, Narok and Laikipia in South rift. Rainfall in the North rift is 1000-1250 mm with clear peaks in April-August and October-December. Low temperatures and adequate rainfall make the region ideal for pyrethrum production.

South rift is the highest pyrethrum producing area and receives 1000-1500 mm of rainfall with peaks in April-May and October and November. The general climate is cold and wet with a mean temperature of 10-15 $^{\circ}$ C characterised by a very long cropping season.

3.2.3 Central

The central regions include Nyandarua, Nyeri, Kiambu and Meru which receive 1000-1500 mm annual rainfall with high peaks between April-May and October-November. The rainfall is bimodal and areas have long cropping season.

Table 5: Rainfall (mm) distribution in major pyrethrum growing areas

Month	Narok Lengetia 2740 m	Nakuru Marindas 2804 m	Marakwet Lelan 2986 m	Uasin Gishu Timboroa 2743 m	Kisii Moromba 1890 m	Nyandarua North Kinangop 2624 m
Jan	44	44	33	21	83	41
Feb	41	45	49	38	96	61
March	70	71	64	72	169	75
April	128	166	110	156	307	144
May	117	127	147	130	255	167
June	85	97	122	118	180	92
July	122	139	195	172	127	71
Aug	196	208	216	211	190	93
Sep	89	97	85	109	191	106
Oct	69	60	95	48	161	99
Nov	71	87	78	51	167	106
Dec	57	64	50	46	132	71
Total	1089	1205	1217	1172	2058	1126

3.3. TEMPERATURE

Temperatures play an important role in the flowering of pyrethrum plants. Some cultivars require low temperatures for flower bud initiation. Such cultivars should not be grown in warm areas lower than 2000m otherwise flower bud initiation may be inhibited. This occurs when mean temperature is above 20°C. Temperatures increase at the onset of the dry period, which cause a decrease in bud initiation and flower yields. The initiation of flower heads requires low temperatures of less than 18 °C for 6 weeks. Temperatures also have a significant effect on pyrethrin content. During the wet season, when temperatures are generally low pyrethrin concentration is high while the reverse occurs during the dry season. Most pyrethrum cultivars have lower pyrethrin content at low altitudes, where the temperatures are relatively high, than at high altitude areas where the temperatures are low.

3.4 ALTITUDE

Suitable pyrethrum growing areas are above 2000 m (table 6). There is a definite correlation between the altitude, temperature and pyrethrin concentration in the flower. Altitudes below 1700 m are warm and plants result into vigorous vegetative growth with no significant flowering. Pyrethrum blooms well from 1700 m and pyrethrin content is higher in high altitude.

Table 6: Mean rainfall, temperatures and altitude in main growing areas

Nyanza			
Area	**Altitude (m)**	**Mean temperature (°C)**	**Annual rainfall (mm)**
Kisii station	1680	16-18	2677
Moromba	1890	15-18	2056
Nyanturago	1680	16-18	2089
Nyathegogi	1620	16-18	1826
North Rift			
Kaptagat	2438	12-15	1261
Timboroa	2743	10-15	1171
Lelan	2986	10-13	1242
South Rift			
Mau Summit	2537	10-15	1025
Marindas	2800	10-13	1209
Londiani	2317	12-18	1182
Kijabe Hill	2537	12-18	1108
Lengetia	2740	10-15	1054
Central and Eastern			
Sasumua	2475	10-15	1620
Ol Joro Orok	2371	13-15	977
North Kinangop	2624	10-15	1123

CHAPTER FOUR

4.0. GROWING PYRETHRUM

4.1. LAND PREPARATION

Pyrethrum is a perennial crop with an economic lifespan of 3 years in the field. Land preparations should aim at controlling perennial weeds such as the Couch grass (*Digitaria scalarum*), Kikuyu grass *(Pennisetum clandestinum*), Star grass (*Cynodon dactylon*) and Sorrel (*Oxalis latifolia*). These weeds are difficult to eradicate later without destroying the established crop. Ploughing should be deep enough to allow deep root growth and establishment, and facilitate water infiltration. Ploughing should be done when the ground is not too wet to cause soil pulverisation or create a hardpan, and subsequent soil erosion.

Seedbed preparation on small fields can be done by digging and breaking up the clods using a hoe, and breaking up the remaining soil until an even minimum depth 15-30 cm is attained. For large fields two ploughings using a tractor or animal drawn disc or mould board plough each followed by one harrowing with a disc harrow give a fine seedbed in a new field. On very rough ground the uses of tine harrow for a final operation may be necessary to level out the soil and remove trash. On fallow land, one ploughing followed by one harrowing give a fine seedbed.

4.2. PLANT PROPAGATION

Pyrethrum can be propagated through splitting of mature plants, seed or tissue-culture. Pyrethrum clones are mainly propagated vegetatively through splits or tissue-culture, while pyrethrum varieties are propagated through seed. A clone is a group of plant population obtained by continuous vegetative propagation of single plant and therefore the entire population is genetically homogeneous and has similar growth characteristics. A variety is a heterogeneous plant population obtained from seed produced through the hybridisation of 2 or more clones and consequently plants in a variety population are different in all growth aspects.

4.2.1. Field Establishment

4.2.1.1. Clonal material

The preparation of splits from mature plants is important for establishment of clonal material. Planting materials from old and woody plants are trimmed at the level of the topmost leaves. Long roots should be trimmed to fit into the planting holes. Splits from growing plants have no old flower stalk or excessive root system. Such plants separate up easily into splits with little damage to the root system. Ensure that the split has adequate roots for establishment. During planting, the whole root system should fit vertically in the planting hole. Roots which bend up in the hole results into poor stand establishment and large quantities of materials required to fill empty gaps. Splits should be planted as soon as they are ready, preferably during the same day plants are uprooted. The splits should be dipped in a fungicide solution to get rid of possible disease infection before planting.

During planting, a split should be held against the vertical wall of the hole, ensuring it is neither too high nor too low in the hole, or at the same level it was in the soil before uprooting (figure 4). Deep planting causes the crown to rot while shallow planting exposes the root, causing it to dry.

(a)   (b)

Figure 4: plant propagation (a) pyrethrum split planting material (b) farmer propagating with splits

The soil should be filled gently with roots hanging downwards and firmed gently around the roots, making it impossible to pull the plant from the soil with a gentle pull. Firming of the soils ensures moisture retention, ensuring good establishment.

If there are no rains, irrigation is necessary to supplement soil moisture. The newly established field should be hand-weeded in the 1st month, by pulling weeds gently to avoid uprooting new plants or damaging roots. A forked implement is ideal for weeding. Using a cutting implement like a hand hoe damages developing root system. Flower buds, which appear during the 1st 3 months after planting, should be destalked manually or with a pair of scissors to promote vegetative growth.

Planting of variety seedlings should be done at the end of rains. Planting seedlings at the onset of the long rains causes flooding and burying. Heavy downpours wash away seedlings. During planting, the young seedling should be held gently and roots placed downwards in the small planting hole. Soil should be gradually filled and firmed up to the level the seedling was in the nursery. There would be good establishment and less need for gapping, if there is a slight shower of rain or irrigation after planting. However, if planting is done during relatively dry conditions, desiccation of the young seedlings might occur and lead to death of the seedlings.

Weeding should be done with forked implements at least once a month after establishment until the crop has fully covered the ground. Gradual removal of early flowering buds should be done to encourage vegetative growth and allow late maturing plants to catch with the rest. Flowers are usually ready for picking in the 5th - 6th month after planting.

After 2-3 months, plants should be earthed-up to encourage tiller production and development. A newly established pyrethrum field produces the 1st flowers in the 3rd or 4th month after planting.

4.2.1.2. Seed material

Seedlings should be transplanted in a commercial field after 5-6 months in the nursery. The seedling is about 15cm tall and has a well-established root system. Each seedling is an individual plant and consequently there is no splitting of materials. Clusters of seedlings are uprooted from the nursery bed and carefully separated into individual plants.

There are 2 methods recommended for planting seedlings: dig large holes and fill it with soil. Then make a small hole large enough to take a root system with 2 fingers or prepare a fine seed-bed and make a small hole using a stick or a blunt implement large enough to take a seedling. This is a faster method and demands less labour.

Advantage of clonal plants or seeds

Clones

- They flower in 3 - 4 months while seedlings flower in 6 months.
- They have higher yields in the 1st year.
- Plants have uniform flowering while seedlings vary.
- Clonal material can be used continuously to replant new fields while seedlings planted in the fields must be discarded after the 4th year.

Seed

- It is cheaper to establish.
- They have undisturbed root system which establish more easily.
- Fields established with seedlings last longer with less decrease in flower yields .
- Owing to their varied genotypes, seedling plantations resist disease epidemics better than clonal fields.

4.3. NURSERY MANAGEMENT

The Pyrethrum Board of Kenya propagates clones and varieties through a network of large-scale and maintains seeds and clonal material in growing areas. The nurseries multiply new elite cultivars. Several steps need to be adhered to establish successful nurseries.

4.3.1. Seedling nursery

The procedure for the preparation and care of a seed nursery is as follows: -

- Make 1.5-2.0 m raised beds of convenient length
- Make straight furrows 15 cm apart and about 1.25 cm deep using a stick
- Place seeds in furrows at a rate of 10 seeds spread evenly on 2.5 cm length of the furrow
- Do not cover the seed with soil
- Cover the seedbed with seedless dry grass and water thoroughly
- Germination takes 10-18 days, after which the grass mulch should be removed in stages to allow the tender seedlings to acclimatise and become strong
- Topdress with CAN (26%N) at 400 kgha^{-1} 3 months after germination

- Seedlings should be transplanted after 4-5 months. Seed should therefore, be sown in October, in order to transplant the seedlings to the field at the onset of the long rains in March or April the following year.

Figure 5: pyrethrum seed beds

Seedlings remain in nursery beds for 4-5 months before transplanting to the field. Any pyrethrum crop that has been established from seed or seedlings should be ploughed back or discarded after 4 years to avoid loss of hybrid vigour in subsequent generations. Such plants are not suitable for replanting onto a new field. Farmers should obtain fresh seed and raise them in nurseries in order to have sufficient seedlings ready for transplanting once the old crop is removed from the field.

(a)   (b)

Figure 6: planting materials nurseries (a) young pyrethrum seedling in beds (b) clonal material multiplication

4.3.2. Clonal nursery

The procedure for the preparation and care of a clonal nursery is as follows:

- Make raised beds of about one metre wide and of a convenient length. Make sufficiently deep holes 15 cm between the lines and 15 cm between the holes in each line
- Split up the clonal material while still fresh before transplanting
- Apply Triple Super phosphate fertiliser at a rate of one teaspoonful per planting hole Also apply a nematicide if necessary to control existing nematodes
- Place a single split into each planting hole making sure the roots are straight (figure 6)
- Press the soil against the roots and ensure the soil reaches just above the root level
- Water the nursery as required
- Plants are ready for splitting and replanting in a commercial field after 3 months.

4.3.3. Rapid multiplication through biotechnology: Tissue culture

Tissue-culture enables mass production of clean, true to type and disease-free planting material. Commercial application of biotechnology in propagation started in 1987. The procedure consists of 5 stages; shoot initiation, shoot multiplication, shoot rooting, acclimatisation and transplanting in the field.

Shoot initiation

Meristems intended for propagation are selected from genetically superior plants with high flower yield and pyrethrins content. Young plants, with actively growing shoots, which have not reached the reproductive phase, are selected and surface sterilized. All auxiliary buds which are 0.5 mm are removed and inoculated in a basal media containing organic salts and low levels of growth hormones and transferred to a well lighted growth room under 3000 lux of light and 18 -22°C to enable the shoots to elongate.. After 10 days the auxiliary bud will have elongated to 3-5cm.

Shoot multiplication

The elongated auxiliary bud undergoes rapid cell multiplication resulting to a mass of shoots. It is the most critical of all the stages and media preparation should be done carefully. Elongated shoots are transferred into a shoot multiplication media with higher levels of growth hormones. The shoots produce multiple secondary shoots and a single shoot may produce 150-200 secondary shoots. After 4 weeks the shoot cluster is removed from the culture jars and shoots separated under sterile conditions. Separated shoots can either be recycled to produce more secondary shoots or advanced to the rooting stage.

Rooting

The mass of shoots obtained in the shoot multiplication are removed from the culture jars and carefully separated into individual shoots. Shoots are consequently inoculated into the basal media with a rooting hormone. Shoots elongate to 8-10 cm and apart from producing roots they may also produce 4-5 rooted secondary shoots. This stage lasts 4 weeks.

(a)

(b)

(c)

(d)

Figure 7: tissue culture propagation (a) shoot initiation (b) shoot multiplication (c) rooting (d) acclimatization

Acclimatization

Rooted shoots are removed from the culture jars and washed in ordinary water to remove media adhering to them. They are then dipped into a fungicide solution to control fungal infections. The seedlings are then planted in 250 ml pots with wet sterilized soil. Potted plants are covered with polythene paper bags in the green-house to maintain high humidity levels of 70-80% for 4 weeks. The polythene papers are half opened after one week to allow air circulation for 2 weeks before the whole polythene paper is finally removed. Acclimatized seedlings are consequently taken out and put under shade for further hardening for another 3 weeks before taking into the field for cultivation. The whole process takes 5-6 weeks.

Transplanting in the field

The young acclimatised plants are ready for planting in a commercial field directly for flower production or transferred to a nursery for further multiplication. Cultural practices such as fertilisation, irrigation, weeding and destalking should be carried out. Young seedlings are disbudded and destalked fortnightly to promote vigorous vegetative growth.

Advantages of tissue-culture over conventional method

- The *in vitro* technique produces up to 1000 plants from shoot in 3 months with no indication of mutations or abnormalities in the plants In comparison with the conventional methods of propagation which yields 5-10 splits from one plant in 6 months
- The technique cleans planting materials and thus reducing the possibility of diseases and pests distribution from nurseries to farmers fields
- Plants propagated through this method have a high establishment rate, more splits per plant, higher tiller production, large bush diameter, more vigorous vegetative growth and higher flower yields
- Rapid multiplication has been used to produce mother clones for production of commercial pyrethrum seeds
- It has been utilised by breeders to reduce pyrethrum-breeding cycle from 9 -11 years to 3 years.

4.4. SELECTION OF PLANTING MATERIAL

4.4.1. Clones

In vegetative propagation, it is advisable to use young vigorous plants for splits. Splits from old and woody plants do not establish well. During selection of planting materials, the following plant types should not be used:

- Blind plants. These are plants that do not produce flowers
- Small, weakly plants which are partially dead. Such plants give poor splits and in most cases are likely to be infested with pests or infected with disease
- Diseased plants should not be split for planting and should always be destroyed.

4.4.2. Varieties

Three commercial varieties are recommended for use in different AEZ. The varieties K218 and K235 are recommended for the low altitude areas starting from 1700 -2200 m while variety P4 is recommended for high altitude areas above 2200 m. It is appropriate to use certified seed from a reputable source to establish a commercial field. Normally, 250 g of seed will give enough seedlings to plant 0.5 ha.

Transplanting

Transplanting in the field should be done early in the rainy season preferably after 2 weeks. High flower yields depend on good plant establishment at 1st planting. This allows quick development of young roots and tillers. Selected plants should be uprooted using a fork jembe, not by hand, to avoid root damage. Soon after uprooting, the flowering stems should be cut off using a sharp knife, and the roots trimmed to a convenient size. Such plants can be split up for planting. The number of splits obtained from a single plant depends on the plant size, the size of splits and the genetic characteristic of the clone.

On average, 5-10 splits can be obtained from a fully-grown plant. Each split should retain much of the original root system. Splits with an insufficient root system establish poorly.

Planting holes should be dug using a panga, a small jembe or a pointed stick, depending on the type of planting materials to be transplanted. Splits will often require bigger holes than the seedlings. Planting holes should be spaced at 60 cm between the rows and 30 cm within rows. This spacing gives a plant population of about 55,000 ha^{-1}. Triple super phosphate fertiliser should be mixed well with soil and applied at a rate of one teaspoonful per planting hole before planting.

4.4.3. Sources of planting material

4.4.3.1. Seed

Seed may be obtained from Pyrethrum Board of Kenya (PBK), Nakuru, Eldoret, Kisii or Nyahururu.

4.4.3.2. Clonal material

Clonal materials may be obtained through the Local Field Officer from any of PBK nurseries in Nakuru, Kericho, Kisii, Nyandarua, Meru, Nandi and Uasin Gishu Districts.

4.5. BREEDING

The aim of pyrethrum breeding programme is to develop clones and varieties with high flower yields and high pyrethrins content, resistance to diseases and pests, resistance to lodging, desirable extract qualities and adapted to different AEZ. Two breeding programmes, clonal and varietal are used to improve pyrethrum.

4.5.1. Clonal breeding

The clonal breeding procedure starts with the selection of superior single plants from a genotypically variable population obtained from a commercial variety. The process consists of single plant selection, single line observation, screening, replicated yield evaluation and adaptation stages that last 9 years (Table 6).

Single plant selection

Commercial seed is sown in a nursery to raise seedlings. Individual heterozygous seedlings are consequently planted in a selection field. Outstanding plants in flowering, non-lodging, vigour and health are selected at the end of the year to advance to single line selection.

Single line selection

Outstanding genotypes from the selection field are split and planted in single rows of 12 plants each and compared with a standard clone (control). Each selected plant must produce more than 12 splits. Further observations are carried out on flowering, growth vigour, ability to split and establish, resistance to diseases and pests and resistance to lodging. Superior genotypes are selected based on these phenotypic criteria before screening.

Screening

Selected genotypes are split and planted in single plots and compared with a commercial clone. Each line is assessed for flowering, pyrethrins content, vigour, ability to split and establish, resistance to diseases and pests and resistance to lodging. Genotypes which are better than the commercial clones are selected and advanced to the replicated yield stage. Superior genotypes are selected based on their phenotypic and genotypic characteristics.

Replicated yield

The selected lines are planted in a replicated yield trial where the genotypes are statistically analysed against a control clone for flower yield, pyrethrins content, ability to split and establish, growth vigour, resistance to pests and diseases, and resistance to lodging. Superior lines based on both genotypic and phenotypic traits are advanced to the adaptability test after 2 years.

Adaptability

All selected genotypes are planted in an adaptability trial in more than 4 AEZ to determine the effect of different environments on their performance. This stage lasts about 3 years. The genotypes are statistically analysed for performance based on flower yield, pyrethrins content, establishment, vigour, resistance to lodging, resistance to pests and diseases. Based on these tests, specific genotypes are recommended for commercial cultivation in specific areas.

4.5.2. Varietal breeding

Four stages; topcross seed production, topcross progeny yield evaluation, diallel crosses and variety yield evaluation are used to develop superior varieties in pyrethrum.

Topcross seed production

Several phenotypically and genotypically outstanding clones from the replicated yield stage of the clonal breeding are planted and allowed to topcross with a pollinator and determine their breeding value. Usually a commercial variety is used as a pollinator. The clones are planted in single rows after every 4 rows of the pollinator. Seeds from each clone are harvested separately and sown in a nursery to raise seedlings for use in the topcross progeny yield evaluation stage.

Topcross progeny yield evaluation

The progenies from the topcross seed production stage are planted to determine the breeding value or the general combining ability (GCA) of the mother clones and the pollinator. The progenies are assessed against a commercial variety as a control for flower production, pyrethrins content, disease/pest/lodging resistance, and plant vigour. Both phenotypic and genotypic traits are utilised to eliminate weak and unproductive progenies.

Diallel crosses

Clones which have high general combining abilities for flower yield, pyrethrin content and other desirable traits of economic importance are planted in double or polycrosses in isolated plots to determine their specific combining ability. The results help to predict the performance of the hybrids or crosses in the field. The seeds obtained from the crosses are used in the variety yield evaluation stage.

Variety yield evaluation

Seeds from the diallel crosses are sown in the nursery and used to plant a variety yield evaluation trial. Each variety is statistically evaluated for flower yield and pyrethrins yield for more than 4 AEZ against a commercial variety. The effect of the genotype environment interaction is reflected in the performance of these varieties and based on these tests superior genotypes are selected for commercial cultivation in specific AEZ.

Inter-relationship between clonal and varietal breeding programmes

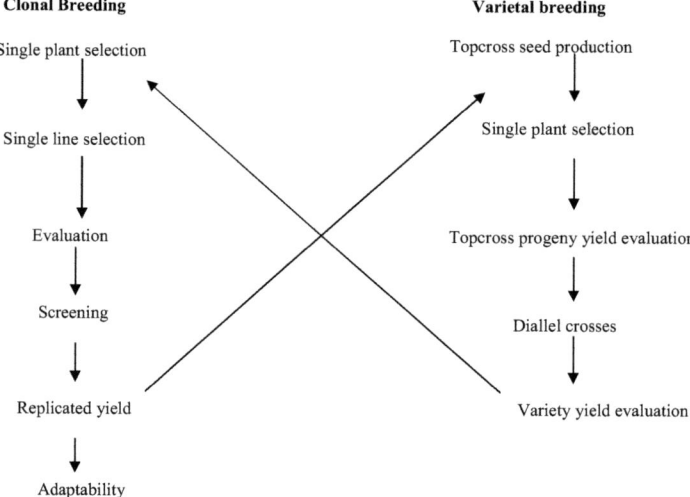

Figure 8: Inter-relationship between clonal and varietal breeding programmes

Table 7: pyrethrum breeding selection criteria

Selection Stage	Selection Criteria	Breeding emphasis	Period of test (years)
Selection	• Resistance to lodging • Resistance to diseases and pests attack • Plant height • Bush diameter • Flower size • Flower yield	• Utilise phenotypic traits to eliminate weak and undesirable cultivars	1
Single line selection	• Resistance to lodging • Disease/pest resistance • Bush diameter • Nematode infestation • Ability to split • Ability to establish • Flower yield • Growth vigour	• Utilise phenotypic traits to eliminate weak clones with poor growth characteristics.	1
Screening	• Resistance to lodging • Resistance to diseases and pests • Bush diameter • Resistance to nematodes • Ability to establish • Vigorous growth • Flower yield • Pyrethrins content • Pyrethrins yield per unit area.	• Utilise phenotypic and genotypic characteristics to eliminate weak clones and select healthy vigorous upright disease resistant clones with high flower yields.	2
Replicated yield	• Flower yield • Pyrethrins content • Pyrethrins yield per unit area.	• Utilise genotypic traits with high heritability values to identify superior cultivars in one or more sites	1-2
Adaptability	• Flower yield • Pyrethrins content • Pyrethrins yield per unit area • Multi-location adaptability.	• Utilise genotypic environment interaction to select superior clones in all aspects of production for commercial cultivation in wide agro-ecological zones	2-3

4.6. FERTILISER USE

Pyrethrum is grown mainly in the highlands at between 1500 and 3000 m. These areas have soils that differ greatly in nutrient composition, texture and structure. However, it thrives best in deep, fertile and well-drained loamy soils of volcanic origin. Such soils are rich in nutrients and retain moisture during dry spells. In marshy and poorly drained soils, pyrethrum is planted on ridges along contour lines.

Fertiliser use show variable responses. In Kinangop, Marindas, Mau-Summit and Kipkabus, phosphorous increased flower yield by 20 -30 % while in Molo, Ol Joro-Orok, Limuru and Keroka there was no positive response, despite being deficient in phosphorous. Various studies show that nitrogen and phosphorous are essential elements for growth. Due to different nutrient status, different areas need specific fertiliser recommendation. Triple Super Phosphate at 200 kgha^{-1} is recommended for pyrethrum at planting.

However, due to decreasing farm sizes, it has become difficult to follow rotational schedules recommended and practised in the past. Also, with the introduction of new cultivars, fertiliser requirement should be reviewed regularly. Soil analysis results of major growing areas indicate that phosphorous is below 20 ppm (Table 8). This implies that phosphorous must be added to promote plant growth. In Sotik, organic carbon and nitrogen were deficient. Sodium was also very low in all the areas sampled.

Table 8: Soil analysis results of different pyrethrum growing areas

Element/ Factor	Site						
	Molo 2500m	M/Summit 2537m	Maji Mazuri 1620m	Sotik 1590m	Nyosia 1600m	Kisii 1680m	Threshold Level
Na m.e.(%)	0.39	0.21	0.18	0.14	0.14	0.22	0.4
K m.e.(%)	1.16	1.02	0.72	0.50	1.12	0.86	0.4 - 0.6
Ca m.e.(%)	7.14	7.80	2.4	1.20	2.0	3.6	2 - 10
Mg m.e.(%)	2.74	1.87	2.1	1.60	2.3	2.3	1.0 - 3.0
Mn m.e.(%)	0.88	1.05	1.23	1.18	0.95	0.80	0.2 - 2.0
P p.p.m	13.0	8.7	14.0	12.0	20.0	16.0	20 - 80
N (%)	0.24	0.21	0.27	0.16	0.29	0.21	0.2 - 04
C (%)	2.03	2.06	2.76	1.58	2.41	1.94	2 - 4
pH	5.60	6.17	5.60	4.80	5.20	5.60	5.6 - 6.20

m.e - Milli equivalents

p.p.m - parts per million

4.6.1. The effect of soil pH on nutrient availability

The soil pH affects the availability of various nutrients necessary for plant growth. In neutral or alkaline soils, active sites should be calcium (Ca^{+2}) or magnesium (Mg^{+2}). In acidic soils, active positions are taken up by aluminium (Al^{+3}) and Ferrous (Fe^{+2}) ions.

Nitrogen availability is highest between pH 6 and 8 mainly due to increase of mineralising and fixing microbes at this pH range. Microbes do not survive in acidic or alkaline soils. The phosphorous compound is readily available for plant growth at between pH 6.5 and 7.5.

Below pH 6.5, phosphorous combines with aluminium and iron to form insoluble and unavailable compounds. Above pH 7.5, phosphorous combines with calcium and magnesium to form insoluble compounds. Potassium, calcium and magnesium are available in alkaline soils. However, as the soil becomes more acidic, the nutrients become less available. Iron and manganese availability increases with increasing acidity because of the increasing solubility of their complex compounds in acidic conditions. Boron, copper and zinc are easily leached and can be deficient in highly leached acidic soils. In alkaline soils boron, copper and silica form insoluble compounds and hence they are unavailable for plant growth.

4.6.2. Soil amendments

Soil amendment compounds are used to alter the physical or chemical property of a specific soil type. The objective is to maintain adequate amounts of phosphorous concentration in the soil in both acidic and alkaline soils.

Sulphur and aluminium compounds are added to amend alkaline soils mostly using Gypsum. The acidic condition is corrected by adding calcium carbonate, Ca_2Co_3 (limestone) or magnesium carbonate, Mg_2Co_3 (Magmax).

4.7. WEED CONTROL

Weeds are a major menace in pyrethrum cultivation and can smother most of the plants (figure 9). In Kenya weed control is done manually and is therefore labour intensive. The ecological requirements of pyrethrum favour the growth of many weeds because pyrethrum does not fully cover the ground. Weeds continually compete with the crop for water, light, nutrients and space. However, different weeds have been identified in pyrethrum growing areas (Table 9).

4.7.1. Manual weeding

Pyrethrum fields should be weeded at 8 week intervals. In more fertile areas with heavy rainfall, additional weeding may be required because weeds establish faster. Newly established fields should be weeded at 4 week intervals.

(a)

(b)

Figure 9: Weed management (a) weedy field (b) clean field

Weeding should be carried out in dry weather to suppress regrowth using fork jembe to minimise damage to pyrethrum root system. The soil should be drawn towards the pyrethrum plants in order to promote the development of tillers and increase water retention around the roots. Weeding, especially of annuals, should be done before weeds reach the flowering stage to get rid of the weed population and to eradicate weeds before they produce propagate or enforce their regenerative capacity.

4.7.2. Preventive weed control

Some herbicides effectively control most weeds with no harmful effects on the crop. Venzar applied at the rate of 1.25 kgha^{-1} will effectively control most weeds for 3-4 months. It should be applied before the field is infested so that it helps in maintaining a weed free situation.

The following methods can be adopted to prevent the introduction and infestation of weeds:

- Use of clean planting materials
- Avoid use of fresh or partially decomposed organic manure
- Do not use organic mulches containing weed seeds
- Livestock should be kept off clean fields as they carry weed propagates in fur, fleece, hooves or droppings
- Use clean farm implements and machinery
- Avoid using irrigation water contaminated with weeds, seeds and other weed propagation
- Keep the field bunds, field channels, boundary bunds, fence lines or head lands free from weeds
- Follow legal and quarantine measures.

4.8. PESTS AND DISEASES

4.8.1. Pests

4.8.1.1. Pyrethrum thrips *(Thrips nigropilosus)*

The pyrethrum thrips, also known as leaf thrips are yellow, minute insects 2-3 mm long and mainly infest pyrethrum leaves and stem. Few thrips, Thrips live on pyrethrum in small numbers during the wet season causing no economic damage. Thrips population increases rapidly during the dry season and starts damaging flowers. Infestation is evident by dirty silvery patches on the leaves, which start to dry. Severe infestation results in plant death.

Egg   Nymph   Pupa

Figure 10: Developmental stages of thrips

Table 9: Common weeds in pyrethrum fields

Common Name	Scientific Name	Molo (2500 m)	Oljoro Orok (2200 m)	Limuru (2300 m)
		Weed density/m^2		
Chickweed	*Stellaria media*	9.8	0.17	5.3
Goose foot/Fathen	*Chenopodium album*	1.0	0.17	0.17
Rape	*Brassicai napus*	0.6	0.50	0.5
Strapwort	*Corrigiolai littorallisii*	3.6	0.17	3.0
Nutgrass/water grass	*Cyperusi rotundus*	0.8	-	-
Kikuyu grass	*Pennisetum clandestinum*	5.8	0.50	-
Sparrey	*Spergula arvensis*	15.3	20.83	2.0
Black jack	*Bidens pilosa*	0.17	-	0.3
Pigweed	*Amaran thus hybridus*	1.17	0.17	0.3
Oxalis	*Oxalis latifolia*	1.8	-	-
Galant soldier	*Galinsoga parviflora*	11.17	0.17	3.8
-	*Conocopus didymus*	23.3	-	-
Mexican marigold	*Tagetes minute*	0.6	-	4.1
-	*Erucastrum arabicumi*	13.17	-	-
-	*Medicago sativa*	0.5	-	-
Goose grass	*Galium spuriumi*	0.17	-	-
Wildfinger millet	*Eleusine indica*	0.3	-	11.8
Couch grass	*Digitaria scalarum*	0.7	-	-
Spiny sowthistle	*Sonchus asper*	0.5	-	1.0
Shepherd's purse	*Capsella bursapastoria*	0.6	-	-
-	*Desmodium indica*	-	-	-
-	*Eragrostis tenuifolia*	0.17	3.33	3.5
wondering Jew	*Commelina benghalensis*	10.6	-	0.5
Flea bane	*Conyza bonariensis*	10.6	-	0.17
-	*Caylusea abyssimica*	0.3	-	-
Black bindweed	*Polygorium convolvulus*	-	-	0.3
-	*Panicum trichocladum*	-	-	7.0
Love grass or Bristly foxtail	*Setaria verticillata*	-	-	-
Sheep's sorrel	*Rumex acetosella*	-	-	2.17
-	*Bulbostylis Schimperiana*	-	7.50	-
Granesbill	*Geranium arabicum*	-	2.17	-
-		-	0.17	-

Flower thrips *(Thrips tabaci)* also known as onion thrips are closely related to the leaf thrips but they only infest the flower head. They cause a browning of the disc florets and ray florets and the flower may dry up prematurely. In the field, thrips are easily detected by shaking a plant onto a piece of white paper. The economic injury level of thrips infestation is 25 adult thrips per leaf.

Both thrips are easily controlled by spraying using insecticides such as Anthio, Diazinon, Dimethoate, Lebacyid and Metasystox at 25 ml/20 l of water.

4.8.1.2. Green peach aphid *(Myzus persicae)*

The green peach aphids are also minute insects less than 2 mm long. They are mainly found on young leaves and growing points. They sack sap from the plants and their secretion covers the leaves, interfering with photosynthesis. Aphid infestations cause distortion of young shoots and leaves. During dry spells attacked leaves dry up. The aphids should be controlled the same way as thrips.

4.8.1.3. Red spider mites *(Tetranychus ludeni)*

These are minute organisms which live on the leaves of pyrethrum where they form a web. The young mite is yellow and changes to red as it moults and grows. The adult is dark red and oval. Infested leaves turn yellow and die in severe infestations. They are easily controlled using Dimethoate, Anthio, Omite and Metasystox.

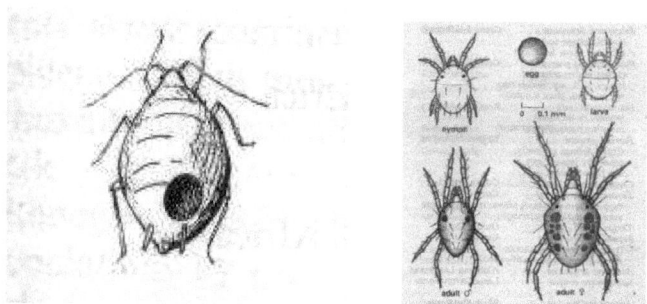

Figure 11: Aphid and mite (a) adult aphid (b) Life cycle of red spider mite

4.8.1.4. Scales *(Coccus sp)*

Scales also infest pyrethrum. They infest the crown and root system feeding on sap and cause plants to wilt in severe attacks. They can be controlled by spraying Diazinon, Dimethoate or any other suitable insecticide as a drench at soil level at 25ml/20L of water.

4.8.1.5. Nematodes

Nematodes are minute micro-organisms and mainly live in soil. Infestation is severe in low altitude areas, which are hot and humid. Severe nematode infestation reduce yields by 20-30 % and cause considerable damage during dry spells. Infested pyrethrum plants appear healthy during rainy season and produce high flower yields but considerable damage occurs during dry spells when plants are vulnerable.

Root knot nematodes *(Melodoigyne halpa)*

This is the most important nematode infesting on pyrethrum. It causes economic damage in all areas, especially those below 2000 m. The nematode is microscopic and feeds on or in roots. It spends most of it's life inside root tissues.

Hatched larvae move freely in the soil to find young growing roots of host plant. Only the females are parasitic. The male remains free, living in the soil. Root knot nematodes lay eggs in cysts in the roots. The mature cysts later burst eggs into surrounding soils. If humidity and temperature conditions are ideal, the eggs hatch into larvae (juveniles), which look for young roots for infestations. Under serious infestation, the whole root system is attacked and the symptoms are characterised by numerous root knots holding masses of eggs.

The roots show stunted growth and develop cysts, which prevent uptake and movement of water and mineral salts. Root knot nematode infestation remains unnoticed as long as growing conditions are favourable for pyrethrum, but yield reduction is realised immediately the plants are under stress in dry weather. Infested plants wilt and die easily in stress conditions. Root knot nematode may combine with root-infecting fungus such as *Fusarium oxysporium* and cause complex root diseases characterised by severe wilting and consequent death of plants.

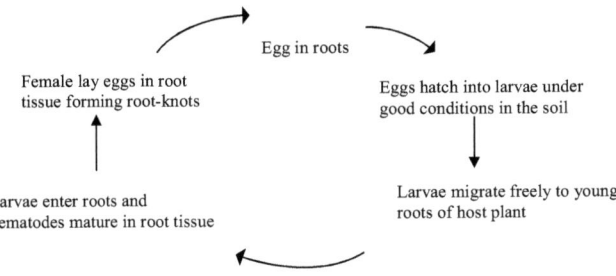

Figure 12: Life cycle of a female *M. hapla* nematode

Root lesion nematodes *(Prantylenchus penetrans)*

The lesion nematodes move freely in soil and feed externally on pyrethrum roots forming lesions on roots. Their damage is restricted to the roots. Lesions facilitate root-rot especially in the dry weather. However, the lesion nematode infestation is not as severe as the root knot infestation and occasionally their infestation is overshadowed by the effects of root knot nematode.

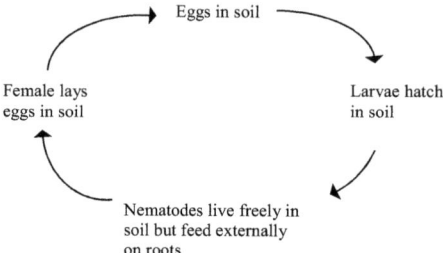

Eggs in soil

Female lays
eggs in soil

Larvae hatch
in soil

Nematodes live freely in
soil but feed externally
on roots

Figure 13: Life cycle of a root lesion nematode

Control

Nematodes are difficult to control and preventive measures are usually preferred. The severity of infestation is greatly influenced by humidity, temperature and rainfall.

(a) Chemical control

Nematicides

Several nematicides effectively control nematodes. These are Furadan, Vydate, Nemacur and Temik which should be applied in planting holes. These are toxic chemicals and should be applied strictly observing manufacturers guidelines. In established fields nematicides are placed around the plants.

Soil fumigants

These chemicals are used to fumigate the soil before planting. The commonly used fumigants are Methyl bromide, Vorlex, Basamid granular and Ethyl Dibromide (EDM). Since the fumigants are expensive and require specialised equipment for application, they are only appropriate and practical in multiplication nurseries.

(b) Cultural control

The following cultural methods can be used to control nematode population in pyrethrum fields.

Crop rotation

After 3-4 years, pyrethrum plants are uprooted and moved to new fields. The field previously under pyrethrum is established with forage grasses such as oats or crops such as wheat, barley or maize for at least 3 years before planting with pyrethrum. Antagonistic plants such as *Tagetes minuta* are often established in the field to control nematodes.

Other cultural methods that reduce injurious effects of nematodes include leaving the field under fallow condition for 3-4 years, deep ploughing and heavy dressing with compost.

Breeding for resistance

Other efforts for reducing nematode effects include incorporating in the breeding programme genetic traits of nematode resistance in pyrethrum cultivars. Some of the commercial clones in use in Kenya have some resistance or tolerance to root knot nematodes.

4.8.2. Diseases

4.8.2.1. Fusarium wilt

The fusarium wilt is the most devastating disease in pyrethrum. It is widely spread in low altitude areas below 1800 m where temperature are relatively high and high altitude areas in Nyanza, Central and Rift Valley provinces (Table 10). It causes stand decline in pyrethrum fields, death of the root system and subsequently the symptoms spread to the crown, tillers and leaves eventually causing wilting and drying of the whole plant (figure 14).

(a)   *(b)*   *(c)*

Figure 14: disease symptoms (a) Wilting plant (b) wilt infested field (c) bud disease

Several fungal species including several species of *fusarium*, *rhizoctonia*, *sclerotinia*, and *ashycochta* have been associated with wilting disease of pyrethrum. Although all these fungus species are pathogenic, *Fusarium oxysporum* is the main causal organism of pyrethrum wilt, causing over 80% of the disease in both low and high altitude areas.

Fusarium wilt is severe in warm areas or during dry hot spells in cooler areas. Under high air and soil temperature coupled with water stress, wilt symptoms including root-rot, crown-rot, chlorosis and drying of leaves and stems, stunting appear in many fields. All infected plants eventually die.

Control

Crop rotation reduces the accumulation of causative organisms in the soil in many areas. Dipping of planting splits in fungicide solutions such as Ridomil, Benlate and Benovap is recommended in *fusarium* prone areas. Incorporation of resistance to *fusarium* wilt in pyrethrum breeding programme is a major selection criteria.

47

Table 10: Percent incidence of fungal and *fusarium* species in 5 pyrethrum growing areas in Kenya

FUNGI	Oljoro Orok 2200m	Limuru 2300m	Molo 2500m	Mwongoris 1690m	Kisii 1680m
Fusarium	95.3	91.7	90.4	92.2	87.7
Rhizoctonia	1.3	2.6	2.3	1.8	2.3
Sclerotium	-	-	-	-	0.5
Sclerotinia	-	-	-	-	5.9
Ascochyta	0.7	-	1.8	-	-
Epicoccum	-	3.6	2.3	-	-
Mycelia sterallia	-	-	1.4	-	2.3
Phomai	-	-	1.8	2.5	3.2
Pythium	-	-	-	1.3	-
Alternaria	-	-	-	1.3	-
Others	2.2	2.1	0.4	0.9	2.3
Fusarium isolate					
Fusarium oxysporum	93.8	90.0	80.0	85.0	81.3
Fusarium solani	0.0	1.3	1.5	2.0	17.5
Fusarium graminearumi	5.0	1.4	0.5	11.4	0.0
Fusarium sambucinum	1.2	1.3	16.5	0.0	1.2
Others	0.0	6.0	2.0	1.6	0.0

4.8.2.2. Bud disease

There are 2 types of bud disease; the true and the false. The true bud disease is caused by a group of fungi *Ramularia bellunensis*, *Alternaria spp*, and *Asychocyta spp*. These fungi attack young buds and flowers. Epidemic outbreaks occur during prolonged foggy and rainy weather conditions. Flower buds dry up and turn brown or purplish grey (figure 14). Growth of the attacked side of a flower is retarded resulting in a deformed bud or flower which bends over to the diseased side.

The false bud disease was previously thought to be physiological because no causal organism had been isolated from the infected plant parts. However, the disease is caused by a flower nematode, *Aphelnchoides ritsema-bosi* which attacks flower buds, which then dry up along with a few millimetres of the flower stalk. The affected bud dies off and bends showing a "shepherd's cool" formation.

Control

There is no control of bud disease its incidence can be reduced by selection of resistant material in breeding programmes. During planting, diseased materials are discarded and at the end of the season plants should be cut back and stalks burned.

4.8.2.3. Root rot

Several fungi such as *Fusarium graminearium, Fusarium samucinum concolor, Fusarium oxysporum, Rhizoctonia spp., Sclerotinia minor, Aschochta spp.,* and *Sclerotinia sclerotiorum* cause root rot. However, only *sclerotinia minor* occurs more frequently.

The fungus spreads by extension of the mycelium to adjacent plants and through soil, especially during splitting. Infected plants wilt slowly and the leaves dry up, followed by death of the plant. In less severe infestations, partial recovery of the plant may occur. Diseased plants are generally weak and easy to uproot due to the rotting of the plant roots.

Control

The root rot disease spreads slowly in established fields. The use of diseased plants for planting increases the spread of the fungi. Only healthy plants should be used for planting new fields.

4.9. CUTTING BACK

At the end of the growing season the flower stalks which remain on the plant gradually dry up (figure 15). Dry flower stalks are cut using a sickle to allow for further growth of new shoots at the onset of rains. Dry stalks should be burnt to ease weeding and reduce the risk of harbouring pests and diseases.

(a)   (b)

(c)   (d)

Figure 15: Cutting back (a) old growth (b) bean legume intercrop after cutback (c) regeneration after cutback (d) blooming field after cut back

49

4.10. CROP ROTATION

After 3-4 years, it is uneconomical to maintain pyrethrum in the field. The crop has reduced vigour and yields due to age, soil exhaustion, diseases and pests. Plants are uprooted in the 4th year and planted in a clean field.

Certified planting materials are obtained from approved nurseries and may be ordered through the Pyrethrum Board's field extension staff or Government agricultural extension service. For rotation purposes, a cereal crop such as maize, wheat, barley or oats should be planted after pyrethrum. Other methods include leaving the field fallow.

Crops such as weeping love grass, Guatemala grass and rapeseed can also be used in rotation with pyrethrum. Such crops reduce nematode population in the soil because they are poor hosts. The aim of crop rotation is to replenish soil fertility, breakdown life cycles of weeds, diseases and pests. During the growth period, pyrethrum at its initial growth stages can also be intercropped with short duration legumes like beans or soya beans (figure 16). This will minimise on weed competition and maximise of land productivity. Farmers should avoid intercrops that can smother the pyrethrum crop like Dolicos, garden peas or maize.

(a)   (b)

Figure 16: Pyrethrum legumes intercropping (a) Beans (b) Dolicos

4.11. FLOWER HARVESTING

The most important factor in pyrethrum production is the quantity of pyrethrins harvested per unit area. Pyrethrin yield is determined by the number of flowers in the plant, flower size, dry matter content of the flower and the pyrethrins content of the flower when dried. Farmers can maximise pyrethrins production through picking flowers at the right time. As indicated earlier (section 2.5.5) pyrethrum flowers develop through 8 stages from the bud to the seed stage. Pyrethrin concentration increases from the bud stage until it reaches a maximum when 3-4 disc florets are open, then gradually declines.

Pyrethrum flowers are selectively picked when the ray florets are horizontal and about 3-4 rows of the disc florets are open (figure 17). At this stage pyrethrin concentration is highest. Flowers with all the disc florets open and those at the early overblown stage may also be picked as they contain appreciable amounts of pyrethrins. Young flowers contain low pyrethrin and if picked in large quantities will lower the pyrethrin content. Flowers picked with excessive moisture are likely to ferment resulting in losses of pyrethrins.

(a)   (b)

Figure 17: (a) A youth picking pyrethrum (b) flowers picked at the right stage

If, picking has to be delayed, all overblown flowers should be picked first since they contain appreciable amounts of pyrethrins. If overblown flowers are left on the plant they reduce the formation of new flower buds. Pyrethrum should not be picked with flower stalks. The best picking is achieved by holding the flower between the first and the second finger and jerking the flower head with the thumb.

4.12. DRYING OF FLOWERS

Harvested flowers should be dried immediately after picking before storage or delivery to the collection centre for onward transport to processing plant. Delay in drying of harvested flowers may lead to deterioration of the flowers due to fermentation/rotting that results to decreased pyrethrins concentration. The ideal small scale method of drying flowers, is by sun drying which is cheaper, efficient and entails no significant loss of pyrethrins. Artificial drying should be used during cloudy weather and especially when drying large quantities of flowers fast to avoid fermentation. This method is desirable during wet and peak harvest periods.

However artificial methods require close attention and great care because they can cause losses of pyrethrins. When using artificial dryers, the drying air temperature should be kept at a maximum of 60 °C to avoid excessive loss of pyrethrins through overheating. Various drying methods are described below.

4.12.1. Open sun drying

Over 90 % of pyrethrum farmers dry flowers on the ground using gunny bags, kavirondo mats and polythene (figure 18). However, the recommended procedure is first to construct trays, which are raised one metre above the ground to facilitate air circulation. The flowers are placed in such a way that they are not more than 10 cm deep and should be turned over several times in a day to ensure uniform drying. The flowers are dry when 4 out of 5 flower heads shatter easily when squeezed between thumb and fore finger. Under favourable weather conditions, the flowers dry in 3-4 days. Under cloudy conditions, flowers dry in 10-14 days.

(a)   (b)

(c)   (d)

Figure 18: Common pyrethrum flower drying methods (a) using gunny bags (b) drying on polythene sheets (c) and (d) raised meshed trays

4.12.2. Combustion drying

There are many types of dryers. Owing to changing circumstances, the use of artificial dryers has been substantially reduced leaving sun drying as the most commonly used method.

Different fuels such as wood, charcoal, saw dust, heavy or light diesel oil and kerosene have been used as sources of energy. The air is heated then passed through the drying beds by an engine-operated fan. The drying temperatures should be maintained below 80 $^{\circ}$C. The direct passing of combustion products such as smoke discolours the flowers. The optimum drying temperature is 60 $^{\circ}$C.

4.12.3. Flue dryer

Another method of drying is direct passage of heated air from source to drying beds by conventional currents. The flue dryer has a fireplace (furnace) connected to a 23 cm galvanised pipe circulated 3 times on the floor of a drying chamber. The pipe ends into a 1.6 m high chimney, outside the chamber. Heated air circulates in the pipes, thereby heating air inside the chamber.

The heated air rises through the 2 tier beds on perforated floors. The heated air takes up the moisture from the flowers in the drying beds and escapes through vents on top of the front wall. At the bottom of the back wall are inlet vents through which fresh air replaces the one in the drying chamber. The dryer consumes about 150 kg of fresh firewood to dry 500 kg of fresh flowers in 6h. Slow burning wood is extremely efficient in drying

4.12.4. Solar dryer

Solar energy is a vast energy source, but it is diffuse, intermittent and subject to interruption from cloud cover. Collection and utilization of solar heat requires the design of efficient and affordable systems for use in pyrethrum drying. Several factors play an important role in the drying process: moisture content, surface area, product to be dried; air temperature and humidity and convection flows. The pyrethrum solar dryer consists of a 90 cm wide drying chamber of convenient length. A transparent polythene sheeting roof is made 55 cm above the chamber. The sides of the drying chamber are covered with polythene sheeting and a 5 cm vent at the upper side of the chamber allows moisture to escape. The drying bed floor consists of a suitably spaced woven mating such as the Kavirondo mat. This size of the dryer can be used to dry 10-15 kg of fresh flowers. The dryer capacity can be increased by increasing the length while maintaining the width of about 1m to facilitate turning of the flowers (figure 19). However, the depth of the flowers in the dryer should not exceed 5 cm. The drying duration depends on the weather condition. It takes 2-4 days and 5-7 days to dry fresh pyrethrum flowers during sunny weather and medium wet conditions respectively. Solar dryers reduce spillage, contamination, fermentation and increases quality of the flowers.

(a)   (b)

Figure 19: Improved drying (a) small scale solar drier (b) medium scale solar drier

CHAPTER FIVE

PYRETHRUM FLOWER HANDLING, MARKETING AND PROCESSING

5.1. PACKAGING AND STORAGE

After drying to about 13% moisture content at the farm level, dry pyrethrum flowers are packed in well-ventilated gunny bags currently provided by Pyrethrum Board of Kenya. Packaging is done gently to avoid losses of the flowers during transit. Each bag carries 25-30 kg of dry flowers. Once packed, the dried flowers should be sent by the individual farmers or farmer groups to the factory immediately. Prolonged storage causes loss of pyrethrins through degradation and spillage. In case of any incidental delay, flowers in bags are kept in dry, cool and well-ventilated places on raised racks to minimise chances of moisture reabsorption.

5.2. DELIVERY CHANNELS

Small-scale farmers market their produce through local co-operative societies, rural self-help groups or individual licences. In addition, the Pyrethrum Board of Kenya operates several crop collection centres in high density pyrethrum farming areas that are not adequately served by the co-operative societies. At the collection centres, the crop is individually inspected, weighed and receipted. Dry flowers are repacked in specially designated gunny bags, labelled, and delivered to the factory for processing. Each gunny bag bears the identity of the grower showing the consignee's name, address and zone number. The label is securely tied onto each bag, with a second label placed with the flowers inside the bag.

5.3. RECEPTION AT THE FACTORY

On arrival at the reception the flowers are checked for moisture content and quality, weighed and a representative sample drawn from each consignment. A receipt is issued for each delivery indicating the grower's name and address, zone number, number of bags received and the net weight of the delivery. The samples representing each of the deliveries received are forwarded to the laboratory for the determination of the pyrethrin content, on which basis the farmers are paid.

5.4. PAYMENT PROCEDURES

After representative samples have been analysed for pyrethrin content, a monthly statement is issued to each of the growers indicating the weight of pyrethrum flowers, pyrethrins content and the computed monetary value of the delivery. The net value payable after debits or credits such as transport charges, bags issued/returned. The rate of payment is determined by the trend of world market price. At the end of the pool year and once the final trading results for the year have been determined any surplus arising from there, is paid out to eligible growers in form of final pool payment according to the total quantity and pyrethrins content delivered by each licensee during the pool year.

5.5. PYRETHRUM PRODUCTION TRENDS IN KENYA

Pyrethrum in Kenya has been fluctuating between 460 to 18,000 metric ton. This has been due to climate change, soil fertility decline, land subdivision, competition for production resources from other farmers' enterprises and market demands. In 1983, Kenya reached its highest production of 18,000 metric ton (figure 20). The highest production of pyrethrum is achieved between October- December because the crop planted during the long-rains flower during this period. The sunny spells during this period favours flower initiation. The lowest production is experienced during April-June period when old crop is cut back and new fields planted. The expansion of pyrethrum production in Kenya has currently been challenged by competition from other farming enterprises like wheat, barley, maize, stevia, tea, potatoes, and vegetables, and dairy, unpredictable/ unfavorable weather conditions, high factory maintenance costs, low pyrethrins recovery efficiency due to high cost of modern extraction equipments, high energy costs, many operations from farm to factory are manual, hence increasing labour costs and requirements, and continued increase of cost of essential inputs and equipment.

Once pyrethrum flowers are received at Pyrethrum Board of Kenya, they are extracted to get the pale extract. The pyrethrum extracts, which forms the major sales component is mainly exported to insecticide formulators in America, Europe, Asia, and Africa for value addition into various formulations. Formulators of consumer end-use products in USA, Europe, Asia use pyrethrum to formulate products such as ant and cockroach killers, flying insect killers (houseflies), aerosols, foggers, garden sprays, shampoos, and insect repellents among others.

There is a great potential for expanding pyrethrum production in Kenya if the internal issues are resolved gradually with public-private partnership. Land under pyrethrum cultivation can be increased further when private sector entrepreneurs and foreign investors find an investment climate more favourable in the future. In 2000 the world demand of dried pyrethrum flowers was estimated at 20,000 tonnes per year against the current production of about 1,000 tonnes. Pyrethrum continues to be competitive with synthetics where selective toxicity and low environmental hazards are important e.g. in the control of insects in storage, food processing facilities and in pre harvest spray. The global pyrethrum based pesticide market is valued at US $ 100million compared to overall pesticide that is worth US $ 28 million. The high demand for pyrethrum than supply offers well for the pyrethrum future. However, the development of the Kenyan pyrethrum based pesticide industry and it competitiveness in the world has been constrained by lack of adequate supplies of the pyrethrum products to local manufacturers.

The development of the pyrethrum industry in Kenya spans over a period of 80 years since introduction of pyrethrum in 1928 from Europe. But the major issue is whether Kenya can be relied on to produce adequate pyrethrum to meet the growing global demand. Available information indicate that Kenya and other pyrethrum producers cannot meet the global demand for pyrethrum which is overwhelming. As a result, several pyrethrum consuming countries have started alternative research and development programmes for pyrethrins production. Other countries with suitable production conditions such as Australia have invested heavily on production and research. Then for the Kenya pyrethrum industry to develop and flourish it is important to;

- Improve and increase farmer incentives to be able to grow and expand the area under the crop
- Improve availability of quality planting materials
- Increase on farm research and development on pyrethrum crop production and usage or development of new products based on pyrethrum.
- Research and development to be able to develop high yielding clones and varieties, and production packages for different growing areas.
- Improve on agricultural extension services to enable farmers adhere to recommended management practices

- Improvement on the infrastructure (roads, power, water supply and telecommunications) in pyrethrum growing areas
- Provide credit for inputs from PBK or financial institutions
- Improve on price and the marketing arrangements to increase profitability of the enterprise

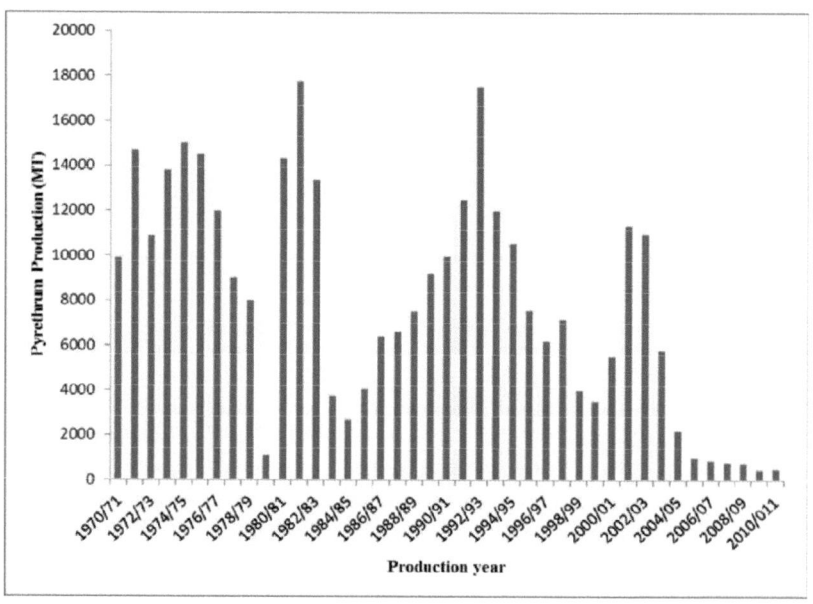

Figure 20:National pyrethrum production figures (MT) in kenya from 1970-2011

In Kenya the global competitiveness of the pyrethrum sub sector contains strengths and weakness and presents both opportunities and threats to the stakeholders. Table Below is a SWOT analysis of the pyrethrum sub sector

Table 11: pyrethrum production strengths, weaknesses, opportunities and threats (SWOT) situational analysis in Kenya.

APVC Component	Situational Analysis				Strategic Partners	Intervention Strategies / Entry Points
	Strengths (S)	Weaknesses (W)	Opportunities (O)	Threats (T)		
Markets & Marketing:						
Policy	Relevant Acts available	Price instability; GOK policy on imports/tax regime discourages domestic manufacture; Lack of competition/ PBK monopoly	Dialogue forums	Conflicting interests; Legislation not appropriate to liberalized economy and globalization	KAPI, Agrochemicals, Johnson Wax Ltd Nairobi, Coil Products	Advocacy by GOK
Channels	Collection centres and Self Help groups available; Established dry flower processing facilities available; Established overseas and domestic market	Late payment to pyrethrum farmers; Poor infrastructure	High extraction capacity at PBK factory;	Cheap imports/ dumping; Decreased production;	Collection centre /self help committees; Ministries Agriculture, Cooperatives, Transport and Roads & Public Works.	Hold joint forums involving all relevant of stakeholders.
Crop Production						
General	Potential for increased pyrethrum production; Traditional position as	Bad farming practice; Unstable pricing trend due to varied pyrethrins analysis;	International trend now favours natural insecticides; Pivotal role	Development of competing producing countries; Increased competition	Ministries of Agriculture, Cooperatives, Transport	Lobby / Consult all stakeholders relevant to the pyrethrum Sub-sub-sector to

57

	worlds largest producer and exporter	Many competing enterprises/lack farmer incentives; Lack of economies of scale; Poor profitability/ net prices to farmers thus constraining availability of products to industry	attractive insecticides in malaria prevention programmes	cheaper synthetics which are available on a regular and consistent basis;	and Roads & Public Works.	address/redress the pertinent constraints..
Policy	Act (CAP 340) available	The Act is outdated.	Act being reviewed; Expansion of usage potential outside traditional	Lack liberalization; Act favours monolithic structure	Select Parliamentary Committee; Ministry of Agriculture	Engage in policy dialogues.
Inputs: Land	Land generally available.	Decreasing land size; Low soil fertility.	Expansion into the non traditional areas.	Social instability due to occasional tribal clashes	Farmers, CBOs, Ministry Lands/Agric.	GoK to stop further land sub-division of unprofitable sizes.
Land preparation	Farm Machinery services are in place already	Services ineffectual.	Tractor hire services at ADC.	Sloppy terrain	ADC; Local tractor hire servies.	Reduce tax on farm implements; Purchase farm implements; Hire tractor services.
Seed/ clonal planting material	Well established network of PBK nurseries	Shortages due to low level of management	Available pure germplasm at the	Disease problems leading to low quality	National Pyrethrum Research Institute, PCPB, KEPHIS,	Inspection by KEPHIS, Activate seed

		National Pyrethrum Research Institute	planting materials of nurseries; Lack of pure clones/varieties		PBK, KEPHIS	regulations.
Agro-chemicals	Knowledge and expertise in downstream value added products; Supply assured	Wide variety of pure chemicals.	Costly (not affordable by most farmers).	Low quality; Threat to the environment.	Agro-dealers, KEPHIS, PBK.	Compliance with ISO rules and KEBS stipulations.
Technology Dissemination:						
Promotion Avenues	System already in place: Use of ASK Shows, and Open/Field Days; PBK extension service	Workshops in farmer training centres; Agriculture classes in secondary/ primary schools (Young farmers and 4K Clubs).	Poor retention of the farming messages by some of the attendants.	Technology redundancy; Farmer fatigue.	MoA, PBK Extension and stakeholders	Make follow-ups to fast track information flow and spread especially among small-scale pyrethrum farmers.
Uptake and Retention Modules	Various (Extension Leaflets, posters, conference proceedings, Electronic / Print media) are available	Training of scientists communication skills.	Too many modules, some of which do not add value.	Possible fatigue on the part of farmers.	MoA, PBK	Craft extension messages in easy language and style.
Research						
Processing	Processing factory available	Linkages avenues can be sought.	Lack of direct modalities of linkage with manufacturers	Fluctuating pyrethrum production over seasons	Manufacturers of pyrethrum products	Invite manufacturers to make their demands based to PBK

Storage	Storage structures available at PBK	PBK have no pyrethrum extract to up store.	Produce from stepped up production.	Redundancy of the structures at PBK	Individual farmers, Collection centres and self help groups	Increase on farmer incentives to increase on productivity to enhance storage
Utilization	There is variation in usage of pyrethrum	Restricted value addition on pyrethrum due to PBK monopolistic structure	Expansion of value addition and utilization potential outside PBK	Competition for the pyrethrum products amongst manufacturers/ consumers	Agrochemicals	Mount questionnaires on the agrochemicals and households on usage.
Crop Protection / Health	National research priorities well defined	Resistance breakdown; Lack of financial resources for research.	Use of IPM strategy; Increased GoK core funding; Grant funds.	Mutations in the form of pathotypes and biotypes; Development of disease/pest chemical resistance	Universities, MoA National Pyrethrum Research Institute, of PBK,	Apply IPM strategy; Mount joint research projects to reduce operational costs.
Crop Husbandry	Recommendation available to farmers	Some options unaffordable by small-scale farmers.	Empowerment of the farmers financially.	Application of sub-optimal rates of inputs.	Credit/ Micro-finance schemes, MoA subsidies,	Avail payments/ funds to farmers in time to encourage adoption of recommendations
Variety Development (Plant Breeding)	Well established plant breeding sub-programme	Some of the breeding techniques outdated;	New integrated/ molecular methods available.	Poor focus on outputs in the sub-programme.	Universities, MoA National Pyrethrum Research Institute, PBK,	Pursue novel approaches (e.g. MAS) to variety development
Germplasm						

Acquisition, Documentation and Conservation	Long experience in Plant Genetic Resources management	Poor maintenance of Germplasm fields	Alternatives sites for storage of duplicates in sub centres.	Internal instability	Universities, MoA, National Pyrethrum Research Institute, PBK, social	Multilocational germplasm establishment to prevent loss of genetic material

5.6. PROCESSING

5.6.1. Grinding

The dried flower heads are coarsely grounded and the grist extracted by percolation with low boiling isoalkanes. The achenes, which contain most of the pyrethrins, are exposed to the extracting solvents for optimal pyrethrin extraction. The solution of extracts is then filtered and evaporated to obtain a viscous liquid called oleoresin. This is the starting material for the refining process. In the extraction of pyrethrins the basic procedure utilised is batch processes, which is based on the separation of pyrethrins from the unwanted vegetable waxes and resins using solvents. Undissolved plant matter from the ground pyrethrum is filtered out and solvent flushed out to produce crude oleoresin, a black, viscous product consisting vegetable matter waxes and pyrethrins.

5.6.2. Refining

The crude oleoresin undergoes further process to remove pigments and foreign plant matter including waxes and resins. The refined product is a clear solution of pyrethrins diluted in refined kerosene to a standard market concentration. The refined extract should have low staining properties and minimum level of inert insolubles. The refined pyrethrins are diluted to a standard concentration of 25% pyrethrins and referred to as technical grade material. This material is consequently blended with synergists, emulsifiers and solvents to produce insecticide concentrations and formulations for consumer product applications.

5.6.3. Properties of pyrethrum

Natural pyrethrum is used for formulation of natural insecticides. The outstanding properties of pyrethrum which enable it to have a wide formulation base are rapid knockdown effect, low mammalian toxicity, biodegradability, repellence/flushing effect and broad spectrum.

(a) Rapid knockdown effect

The most important feature that dictates the selection of pyrethrins as a key formulation component is the rapid insect knockdown it produces on various insect pests. It is a contact insecticide attacking the nervous system of insects immediately and causing knockdown followed by death.

(b) Low mammalian toxicity

The low mammalian toxicity of pyrethrins provides a moderate level for use in and around domestic environments. It is the least toxic of domestic insecticides available in the market.

(c) Biodegradability

Pyrethrum is degraded by the combination of sunlight and air and therefore presents little hazard, which is usually associated with synthetic insecticides. This property makes pyrethrins a safe and environmentally friendly insecticide for use in sensitive areas such as homes, hospitals, factories, stores and on domestic animals and pets.

(d) Repellence/flushing effects

Pyrethrum is a powerful insect repellent and has a strong flushing effect on insects, especially cockroaches.

(e) Broad spectrum

Pyrethrum has a wider spectrum of activity against insect species than synthetic pyrethroids and insects have not developed resistance to it.

5.6.4. Synergism

Synergists are synthetic chemicals with little or no insecticidal activity of their own but enhance the potency of pyrethrins when added in different formulations. The combination of a synergist, which is comparatively inexpensive with the more expensive pyrethrin, enables an equivalent insecticidal effect at a lower cost. Besides increasing the toxicity of pyrethrins to insects, synergists also increase the speed of knockdown and the rate of activation commonly referred to as the "flushing" action, by which certain insects, particularly cockroaches are driven from their hiding places. If one increases the amount of synergist, the biological activity will also increase, but gradually tail off to an optimum level after which it would be uneconomical to add any more synergist.

Synergists, work by preventing the insects destroying the pyrethrins which would otherwise kill it. However, different insect species have various capacity to destroy pyrethrins. For instance, adult mosquitoes have a poor ability to destroy pyrethrins unlike houseflies that have a relatively greater ability to destroy pyrethrins and consequently large quantities of synergists would be required in housefly control formulations.

Besides their biological value many synergists provide useful additional properties of their own in formulations. For instance, Piperonyl butoxide apart from being an excellent solvent helps to prolong the action of pyrethrum for many applications.

5.6.5. Formulations

After processing and refining, pyrethrum is marketed as a standardised extract containing 25% weight/weight (w/w) pyrethrins. This extract is supplied in 3 qualities, oleoresin extract, partially dewaxed extract and pale extract.

Different pyrethrum formulations for control of different of insect species are made from the 3 extracts. These formulations can be categorised into aerosols, repellants, storage powders, pet products, veterinary products, crop sprays and animal feeds.

5.6.5.1. Aerosols

Aerosols are basically pressurised insecticide holding packages designed to dispense known quantities of indicated active ingredients as a space spray. Aerosols can be categorised into 2 main classes, oil-and-water based sprays.

These sprays are formulated using designated pyrethrins concentration and heavier refined mineral oils or emulsifying agent containing water. They consists of household and industrial sprays and have a low absorption rate on the sprayed surfaces and provide a thin residual surface fume with longer residual action. They are used to control flies, mosquitoes, cockroaches and control of storage pests.

5.6.5.2. Pyrethrum powders

Although many synthetic insecticides can be formulated using inert filler material, pyrethrum has few material that can be used to formulate insecticidal powders. Dust fillers with high specific surface areas and good absorbent properties cause pyrethrum to loose its biological activity and potent. The main inorganic fillers used in the formulation of pyrethrum insecticide powders are Talc, B,P,C, Talc 13 and Talc 20. Fillers such as Tacum H, Silice D, Diluex A and Pyraxi can also be used. Materials such as diatomite and clays produce unstable powders and should never be used with pyrethrum. The pyrethrum insecticidal powders are used to control fleas, termites, ants, cockroaches, lice and storage pests.

5.6.5.3. Mosquito coils

Mosquito coils based on pyrethrum are designed to burn for 8 h providing protection from mosquito. They contain 1.3% pyrethrum powder, filler, binder, dye and a fungal agent. When ignited, the slow burning coils produce smoke, which acts as a killing agent and a repellent. Pyrethrins initiate and activate mosquitoes to fly away from the source of the smoke. At a higher dose of pyrethrin concentration, mosquitoes may be knocked down and death may occur.

CHAPTER SIX
COMMERCIAL APPLICATION OF PYRETHRUM BASED PRODUCTS
6.1. AGRICULTURE

The instability of the pyrethrum molecule when exposed to sunlight has limited wide-scale use of pyrethrum-based products in agricultural crops. Successful introduction of cheap and photo stable synthetic pyrethroid insecticides for the control of commercial agricultural insect pests has restricted the use of pyrethrum products for indoor and domestic insect control. Pyrethrum is not cost-effective when used alone in major commercial crops. However, due to its low mammalian toxicity and rapid degradation by ultraviolet radiation, pyrethrum is exempt from the establishment of tolerance rates when applied on various crops, thus allowing it to be used up to and including the day of harvest on all crops. Due to recent improvement in formulation and following the need to use soft insecticides on sensitive crops, pyrethrum has continuously and steadily attracted specific market niches, especially in high valued minor export crops such as green vegetables, fruit trees and ornamental flowers where there are few safe insecticides.

Pyrethrum products could also be used in major crops at lower and more economical rates to control pests susceptible to pyrethrins when mixed with a conventional insecticide. This application utilises the rapid action of pyrethrins on insect population, thus improving control. Pyrethrum could be used to control specific insects and preserve beneficial arthropod insects by utilising low application and insect behavioural patterns. For example, in a blossoming crop, pyrethrins could be applied early in the morning and late in the evening to control pests and allow bees to pollinate the flowers during the day.

In addition pyrethrum is widely used to control various pests, due to its low mammalian toxicity and short residual life which allow its application in numerous situations.

Pyrethrum based insecticides are effective against several horticultural pests such as aphids, sawflies, diamond back moth, thrips and red spider mites which cause economic losses through stunted growth, wilting of leaves and flowers, distortion, abortion and malformation of fruits. Pyrethrum is effective on pests of leaf vegetables, fruit trees, bulbils, tubers, and horticultural cut flowers.

6.2. PUBLIC HEALTH

Pyrethrum products have also been used in the control of different household pests in the public health sector.

6.2.1. Cockroaches

Cockroaches are unclean and unpleasant insects that spread pathogenic bacteria, virus, fungi, protozoa's and worms that cause disease of great economic significance. Cockroach populations increase in warm and moist conditions with adequate food and water. Eggs are laid in batches in pod like cases or capsules. They hatch into nymphs, which pass through 6 nymphal stages by moulting 5-10 times to become adults in 40-50 days. Cockroaches aggregate in dark and concealed places such as wall cracks, cupboards, lavatories, water pipes, sinks, cabinets, door hinges, under and inside cookers, boilers, boiler rooms and ceilings in household kitchens, food processing units and food stores.

Pyrethrum-based formulations are used to flush out cockroaches from their hiding places driving them to come into contact with the insecticide to cause knockdown effect and death. There is a rapid excitatory response of cockroaches to pyrethrins at low concentration, which is used to determine the extent by roach infestation. The flushing action of pyrethrins is effective between 0.01 and 0.05% and it is superior than that by synthetic pyrethroids. Pyrethrum is mixed with piperonyl butoxide at 1:5 ratio to obtain optimal results. A residual compound may be added to a pyrethrum formulation to kill cockroaches that may invade the premises after application.

6.2.2. House fly

The use of synergised pyrethrins aerosol has been used to control houseflies in households, hotels, butcheries, open air markets and refuse dumps. A formulation containing 0.05% pyrethrins and 0.4 piperonyl butoxide proved effective.

6.2.3. Bugs, lice, mites and fleas

The use of synergised pyrethrum with low toxicity effects has been used to control ectoparasites parasiting on man and pets. The ectoparasites are irrational and may cause plague (fleas), pediculosis and typus (lice) on the host. Formulation of pyrethrins at 1:10 has been used to control head and body lice.

6.2.4. Mosquitoes

Different species of mosquitoes transmit diseases of great economic damage to man in the wet humid tropical and subtropical regions. The female anopheles mosquito transmits *Plasmodium falciparum* protozoan, which cause malaria. This is a major killer disease in the tropics.

Other species of mosquito transmit yellow fever, filariasis, encephalitis and other insect-vector-borne diseases. The diseases transmitted by mosquitoes not only causes mortality but also creates cycles in which economic development is affected in lost working time and cost of medication. Each year 300-500 million people get malaria in the world. Of the more than 3 million that die, 50% are children below 5 years old.

Pyrethrum-based aerosols, mosquito coils, lotions, evaporators, chips and treated mosquito nets are used to control and repel mosquitoes in domestic premises.

6.3. LIVESTOCK

6.3.1. Pymarc

Pymarc is a by product of pyrethrum extraction. It is used as a feed supplement for livestock. Pymarc is comparable to other common feeds supplements. Pymarc is mixed with other feeds such as hay, molasses, bran, pollard and daily meal to improve on its palatability. Livestock fed on pymarc have a reduced load of endo-parasites and improved physical appearance. Pymarc contains 13 % protein, 56 % carbohydrates, 23 % fibre, 7 % minerals and 1 % oils on weight basis.

6.3.2. Tick control

Ticks are important livestock pests in Africa, causing enormous losses in milk, meat and hide and skin production. These losses are due to livestock mortality, lowered productivity of survivors due to irritations, damaged hides and predisposition of the animals to bacterial and fungal infections in the exposed wounds caused by tick-bites. Young cattle that survive infection become stunted and may require many months to regain normal growth. Ticks and associated diseases constraint viable livestock industry in the tropics. *Boophillus decoloratus* (Blue tick) cause babesiosis (Red water) in cattle. Another important tick species that cause high mortality in cattle is *Rhipicephalus appendiculatus* (Brown ear tick) which causes East coast fever (ECF). Also *Rhipicephalus evertsi* (Red legged tick) cause anaplasmosis. Several other species of ticks such as *Ixodes Scapulais, Ornithodorus moubata, Amblyomma americum, dermancentor variabilis* and *Rhipicephalus saguineus* also cause damage in cattle (Table 11). A pyrethrum concentrate containing 4% pyrethrin and synergist at 1:5 with piperonyl butoxide and diluted to 0.05% pyrethrins controls several tick species. Weekly spray of pyrethrin-based formulations is as effective as commercial tick spray in combating tick populations under field conditions.

Table 12: Diseases transmitted by ticks

Disease	Tick species		Causal organism	Animal	Site
	Common name	Scientific name			
East coast fever	Brown tick	*Rhipicephalus bursa*	*Theileria parva*	Cattle	
Red water *(Babesiosis)*	Blue tick	*Boophilus decoloratus*	*Babesia* spp	Cattle/sheep	Face, neck, dewlap
Red water *(Babesiosis)*	Brown ear tick	*Rhipicephalus appendiculatus*	*Babesia* spp	Cattle/sheep	Ears, base of horns, eyes, tail brush
Anaplasmosis *(Gall sickness)*	Blue tick	*Boophilus decoloratus*	*Anaplasma marginale*	Cattle/goat/ sheep	Face, neck, dewlap
Heartwater	Bont tick	*Amblyomma maculatum*	*Cowdria ruminantium (Rickettsia)*	Cattle/goat/ Sheep/pigs	Under tail, udder, scrotum, tail brush, heels

6.3.3. Biting flies

Several formulations of pyrethrins are used in sheds, barns, stables, dairies to control biting flies that attack livestock. These flies include stable fly *(Stomoxys calcitrans)*, hornfly *(Haematobia irritans)*, the screw-worm fly *(Cochliomyia macellaria)*, the black blow fly *(Phormiai regina)*, the green bottle fly *(Lucilia sericata)*, the heel fly or warble fly *(Hypoderma lineatum)*, horse flies *(H. bovis)* and deer flies *(Tabanidae)*. Pyrethrum has also been used commercially for wide-scale control of different species of tsetse flies including *Glossina pallipides, Glossina synnertoni* and *Glossina morsitans*. Pyrethrum formulations are also used to control fleas, lice, mites and bedbugs on domestic pets and on human beings. Other diseases transmitted by biting flies are shown in Table 12.

Table 13: Diseases transmitted by biting flies

Disease	Fly vectors	Disease organism	Host	Area prevalent
Malaria	Mosquito (Female Anopheles)	Protozoan (Plasmodium)	Humans Monkeys	Tropics and sub-tropics
Yellow fever	Mosquito (*Aedes aegypti*)	Virus	Humans and Monkeys	Tropics and sub-tropics
Filariasis (Elephantiasis)	Mosquito (Anopheles and Aedes)	Nematode (*Wuchereria bancrofti*)	Humans, dogs, wild carnivores	Tropic and sub-tropics
Break bone fever (Dengue)	Mosquito (Culex and Aedes)	Virus	Humans	Tropics and sub-tropic
Encephalitis	Mosquito (Aedes and Culex)	Virus	Humans	Temperate and tropics
Anthrax	Horseflies (*Tabanus*)	Bacterium (*Bacillus anthracis*)	Ungulates, rodents, humans	Widespread
Sleeping sickness	Tsetse flies *Glossina* spp)	Protozoan (*Trypanosoma*)	Humans	Tropics
Nagana	Tsetse flies (*Glossina* spp.)	Protozoan (*Trypanosoma*)	Domestic and wild mammals	Tropics

The economic importance of biting flies rests primarily on their role as carriers of disease producing organisms of humans and animals. In several cases biting flies represent the sole method of transmission, and the parasite undergoes part of its life cycle in the fly. However, amongst most of the biting flies only the females are bloodsuckers and males and gnats commonly subsist on dew, honeydew, and flower nectars. Most of the diseases transmitted by these fly vectors are under a considerable measure of control, either by vector control, control of the infection via drugs or immunisation.

6.4. STORAGE

Various storage pests attack grains and legumes in storage causing 20-30% loss either directly or indirectly. Directly, storage pests bore narrow tunnels from the surface towards the inside of kernels, destroying the germ and the endosperm which result to reduced weight. Living or dead insect, cast exoskeleton, pupal cases and cocoons, faecal material and persistent odours also contaminate grain. Further the insect activity inevitably results in accumulation of dusts.

Indirectly, the metabolic activity of storage pests increases temperature and moisture in the grain thus increasing the development of moulds and fungi. High temperatures and moisture accumulation in the grain mass reduce their germination capacity. Storage pests also enhance development of disease-causing organisms such as salmonella and moulds that produce aflatoxins to warm blooded animals. Pyrethrum formulation synergised by piperonyl butoxide either in form of powder, fog or aerosols have been used to control and repel storage pests.

6.5. SYNTHETIC PYRETHROIDS

Synthetic pyrethroids are chemical variants of natural occurring pyrethrins. They are mainly manufactured by modifying the benzyl chain of pyrethrin I molecule. Several synthetic pyrethroids with different chemical properties for specific insect species have been developed. Pyrethroids such as permethrin, cypermethrin, deltamenthrin and fenvalerate have been used to control different insect species on agricultural crops. Other compounds such as resmethrin and bioresmethrin have high killing power and low knockdown effect on insects. Others such as tetramethrin and bioallethrin have high early knockdown effect and relatively low killing power. Pyrethroids have less repellence than natural pyrethrins when used in mosquito coils.

The wide-spread use of synthetic pyrethroids in agriculture to control insect pests on field crops is mainly because they remain stable when exposed to air and light. Synthetic pyrethroids are cheap, easily available and are effective on target insect pests. Synthetic pyrethroids contribute about 25% of the agricultural insecticide industry in the world. A lot of research aimed at increasing their effectiveness and lowering their environmental effects and insect resistance. This wide application of synthetic pyrethroids reflects their importance in the agricultural insecticide industry.

Despite these remarkable developments in the formulation of these chemicals, it has continuously eluded researchers to develop chemicals with similar properties to those of the natural pyrethrins in low mammalian toxicity, environmental safety, low insect resistance and rapid and powerful action against insect pests. The major constraints limiting continuous wide-scale use of synthetic pyrethroids is the increased development of resistance in insect population even after a short period of use and their accumulation in the environment.

CHAPTER SEVEN

CONCLUSION AND WAY FORWARD

In Kenya, botanical crops and products remain underutilized mainly because of the lowly explored economics and potency. Currently the most users of the botanicals are the poorly resourced and illiterate farmers who opt for botanicals as an affordable alternative to the costly inorganic chemical and products especially in the field of integrated pest management.

There is need for more exploration in botanicals and organic products, development of proper agronomic packages, creation of awareness/knowledge and promotion on the use of environmentally friendly products if to improve the country's agricultural status and competitiveness in world markets. The above can also boost the country attain its vision 2030 and world millennia goals on environmental conservation. Continuous strategic research in basic and adaptive research in breeding, agronomy and botanical plant protection will yield technological packages and recommendations for different growing areas. Agronomic research on plant population, fertiliser use, weeds control, harvesting and processing require to be developed and recommendations made.

Despite the aggressive competition posed by the synthetic pyrethroids to natural organic products like pyrethrum, demand has remained steady due to their unique properties. There has been a steady shift in demand by consumers to natural insecticidal products in all major segments of the primary and secondary markets in the world due to current health concerns. This trend has increased interest in manufacturers in the production of natural products with an aim of capturing the emerging markets.

Recent developments in the formulation of pyrethrum and its application for control of insect pests under various situations and its renowned properties of safety, non-persistence, environmentally friendly, rapid knockdown and death has made pyrethrins very popular insecticide. New formulations for outdoor use on field crops and domestic animals have opened new areas of use, which was previously ignored or thought impossible. Farmers should be sensitized on the need to start commercial production of other botanical crops and use improved technology packages to increase their yields per unit area. Manufacturers should also be encouraged to initiate use of other botanical plants together with pyrethrum as a basis for the manufacture of agro-chemical products. Even with severe competition from synthetic pyrethroids, demand for natural pyrethrum products will continue to increase as mounting legislation is enforced in consumer countries.

REFERENCES

1. Casida, J.E and G.B, Quistad.1995. Pyrethrum flowers: production, chemistry, toxicology, and uses. Oxford University Press, New York, USA

2. Gichuru, S.P., Ottaro, W.G.M., Ngugi, C.M. and Ikahu, J.M.K. (1990). The use of tissue culture technique for commercial propagation of pyrethrum clones in Kenya. Biotechnology Kenya pg-182-187.

3. Goetyn, R., Kimani, P.M. Ikahu, J.M. and Ngugi, C.W. (2001). Pyrethrum Chrsanthemum cinerariefolium Trv. Bocc). Crop production in Tropicl Africa pp 1141-1148. Edited by Roman H. Raemakers. Directorate General for International C0-operation, Brussels, Belgium.

4. Ikahu, J.M.K. (1988). Phenotypic variation in floral development and pyrethrins content in genetically different clones of pyrethrum. MSc thesis University of Nairobi.

5. Ikahu, J.M.K. and Ngugi, C.W. (1988). Yield assessment of newly developed pyrethrum varieties in different ecological zones in Kenya. Pyrethrum Post Vol.17(1):21-

6. Ikahu, J.M.K. and Ngugi, C.W. (1990). Floral development in some pyrethrum clones and its implication in picking. Pyrethrum Post Vol. 18(1):11-14.

7. Ikahu, J.M.K. Ngugi, C.W. and Maengwe, E.O. (1994). The performance of recommended clones and Kisii clones in different ecological zones in Kenya. Pyrethrum Post Vol. 19(2): 47-53

8. Ngugi, C.W. and Ikahu, J.M.K. (1990). The effect of drying temperature on pyrethrins content in some pyrethrum clones. Pyrethrum Post Vol. 18(1):18-21.

9. Ngugi, C.W. Ikahu, J.M.K. (1989). The response of pyrethrum to phosphorus and nitrogen fertilizers. Pyrethrum Post Vol. 17(2):70-73.

10. Ngugi, C.W. Ikahu, J.M.K. and Gichuru, S.P. (1989). The effects of venzar in weed control in established pyrethrum fields. Pyrethrum Post Vol. 17(2) 52-55).

11. Ngugi, C.W., J.M.K, Ikahu, and G.K, Gathungu. 2006. Pyrethrum production for Agri-business and economic empowerment. Kenya farmer issue 138: Volume 3 pg. 48 – 50.

12. Parlevliet, J.E. & Brewer, J.G. (1971). The botany, agronomy and breeding of pyrethrum, (*Chrysanthemum cinerariaefolium* vis. Report of the Ministry of Agriculture Molo, Kenya.

13. Parlevliet, J.E. and Contant, R.B. (1970). Selection for combining ability in pyrethrum, (*Chrysanthemum cinerariaefolium* Viz. Euphytica 19: 4-11.

14. Tegemeo. 1999. Stakeholders Workshop on the way forward for the Kenyan pyrethrum industry. Proceedings and recommendations, workshop held on 2nd December, 1999 at Kunste Hotel. Tegemeo. Egerton University.

15. Coleman, P. 2012. Guide for Organic Crop Producers. National Organic Program Agricultural Marketing Service

16. Gupta, S. and A. K. Dikshit. 2010. Biopesticides: An eco-friendly approach for pest control. Journal of Biopesticides 3(1 Special Issue) 186 – 188

17. International Centre for Research in Organic Food Systems (ICROFS). 2010. Organic cropping Systems for Vegetable production - product Quality, natural Regulation and Environmental effects (VEG-QURE 2007-2010)

18. Kandpal, V. 2014.Biopesticides. International Journal of Environmental Research and Development, Volume 4, Number 2 (2014), pp. 191-196

19. U.S. Department of Agriculture, Washington, USA. 64p.

20. Zibaee, A. 2011. Arash Zibaee (2011). Botanical Insecticides and Their Effects on Insect Biochemistry and Immunity, Pesticides in the Modern World - Pests Control and Pesticides Exposure and Toxicity Assessment, Dr. Margarita Stoytcheva (Ed.), ISBN: 978-953-307-457-3, InTech, Europe.

21. EPC.2014.   trade   and   production   statistics   on   pyrethrum   industry. http://epckenya.org/images/stories/Publications/pyrethrum%20trade%20and%20production%20statistics .pdf